Different Perspectives in Design Thinking

Editor
Yvonne Eriksson
Professor of Information Design
School of Innovation Design and Engineering
Mälardalen University, Eskilstuna/Västerås, Sweden

CRC Press
Taylor & Francis Group
Boca Raton London New York

CRC Press is an imprint of the
Taylor & Francis Group, an **informa** business

A SCIENCE PUBLISHERS BOOK

Cover illustration courtesy of Gabriela Goldschmidt

First edition published 2022
by CRC Press
6000 Broken Sound Parkway NW, Suite 300, Boca Raton, FL 33487-2742

and by CRC Press
2 Park Square, Milton Park, Abingdon, Oxon, OX14 4RN

© 2022 Taylor & Francis Group, LLC

CRC Press is an imprint of Taylor & Francis Group, LLC

Reasonable efforts have been made to publish reliable data and information, but the author and publisher cannot assume responsibility for the validity of all materials or the consequences of their use. The authors and publishers have attempted to trace the copyright holders of all material reproduced in this publication and apologize to copyright holders if permission to publish in this form has not been obtained. If any copyright material has not been acknowledged please write and let us know so we may rectify in any future reprint.

Except as permitted under U.S. Copyright Law, no part of this book may be reprinted, reproduced, transmitted, or utilized in any form by any electronic, mechanical, or other means, now known or hereafter invented, including photocopying, microfilming, and recording, or in any information storage or retrieval system, without written permission from the publishers.

For permission to photocopy or use material electronically from this work, access www.copyright.com or contact the Copyright Clearance Center, Inc. (CCC), 222 Rosewood Drive, Danvers, MA 01923, 978-750-8400. For works that are not available on CCC please contact mpkbookspermissions@tandf.co.uk

Trademark notice: Product or corporate names may be trademarks or registered trademarks and are used only for identification and explanation without intent to infringe.

Library of Congress Cataloging-in-Publication Data (applied for)

ISBN: 978-0-367-25423-0 (hbk)
ISBN: 978-1-032-12497-1 (pbk)
ISBN: 978-0-429-28937-8 (ebk)

DOI: 10.1201/9780429289378

Typeset in Palatino
by Radiant Productions

Preface

This book is about different perspectives in design thinking. The contents span a broad spectrum of perspectives in design thinking from design methods, design theory to immediate application of design in various contexts. A key theme in the chapters, is the role of design in various contexts and how it affects individuals and groups. The need of a systematic exploration of the role of design and designers in society are problematized by the authors and how the shape of the design affect peoples' life and understanding of the environment. Design is much about communication, communication in the design process and communication with the people and the environment. One of the challenges with design research is to systematically evaluate the design process and its result.

The purpose of the book is to contribute to the area of innovation and design and to elucidate various perspectives that the authors found in design thinking. The book is directed towards a broad audience. It is also intended for students and researchers in design, engineering design and information design and with an aim to present the material in a form that is comprehensible for graduate, advanced undergraduate and Ph.D. students. The book is also designed to attract readers from outside of the immediate research area.

I would like to thank the authors who have contributed to the book and a special thanks to Gabriela Goldschmidt for providing the photo for the cover of the book. The picture showing the masks with their empty eyes illustrates well the relation between what is out there to perceive and what we see and look at, that is the relation between the mental image and the physical environment. But the picture also illustrates that it is always possible to give different perspectives of theories and methods.

I would also like to thank my editor Vijay Primlani for his helpful and very important guidance through the process with the book.

At last, I will thank my partner Judit for her support, during the work with this book.

Yvonne Eriksson

Contents

Preface iii

Introduction vi
Yvonne Eriksson

1. Design Thinking between Myth and Methodology 1
   Petra Badke-Schaub

2. Critical Design and Design Thinking vs. critical design 6
   and design thinking
   Gabriela Goldschmidt

3. Adding Rigor to Advance Design Thinking 21
   Andrew Dillon and Marian Sweeney

4. Situated Design Thinking 41
   Åsa Wikberg Nilsson

5. The Challenge of Designing Meaningful Information 67
   Yvonne Eriksson and Anna-Lena Carlsson

6. Gendered Design Thinking—So-called Logic of People and Things 88
   Åsa Wikberg Nilsson and Yvonne Eriksson

7. On Design Dialogues—Their Roots, Features, and Usage 108
   Ulrika Florin

8. Design Thinking and Designerly Ways of Knowing in 131
   Operational Research Practice
   Christina J Phillips

9. Design Thinking and Welfare Technology: 165
   A Focus on Information Design
   Riika Saurio, Lea Hennala, Satu Pekkarinen and Helinä Melkas

10. Raising Users' Confidence in their own Technology Literacy 202
    as part of the Design Process
    Marie Sjölinder

11. Lessons Learned: A Plea for Curricularizing Design Thinking 220
    in Engineering Education
    Anne Wallisch and *Kristin Paetzold*

Index 245

Introduction

Yvonne Eriksson

Design can be found everywhere in society in the form of systems, services, infrastructure, architecture, artifacts, and tools. The examples are endless. Still, people primarily associate design with beautiful objects, furniture, and fashion—all of which, of course, constitute designed artifacts. But actually, design affects much of our lives, whether the design serves as a support or a hinder to our ability to act or to understand and use products or services. Products and services can be designed to be used intuitively but, oftentimes, the design is divided into parallel pipes that do not correspond with one another and, for this reason, cause problems among people. Therefore, it is necessary to take a holistic perspective. Not only to point towards one or two points, but to see how a change or the development of details will affect a whole system. To do so, different competences need to be involved in the design process, not least during the framing of the problem. We often look for a solution before we have recognized what is causing the problem in the first place. Quite frequently, what needs to be changed is hidden from us at first glance. It is equally challenging to come up with new results in a design that do more than just show a variation of something that already exists. This can be explained by our limited capacity to shake off what we are used to, that which is a part of our heritage and cultural influences.

Design is culturally based. The way products and systems—including infrastructure, services, healthcare and cities—are designed differs from culture to culture. This is based on both value differences between cultures, as well as variations in climate and topography. A change of a product or services doesn't necessarily lead to something better, but it could lead to something different. Design is then not only about creating, but it is also about the ability to analyze current situations and previously existing services before starting to create new solutions. If we cannot evaluate the

---

School of Innovation, Design and Engineering, Division of Information Design, Mälardalen University.

current situation through analysis, it is not possible to assess whether the new product or service represents an improvement or not.

Design can carry many different meanings. It can be a concept, an action, a plan or intention, an outcome or a product. Design is also about communication and research. Design research can encompass research in design or by design, but it can also be about design itself. There are several links between research through design and about design, and cognitive psychology and behavioral psychology. This is because different perspectives on design related research seek to understand how the design process works (what is in the minds of the designers) and how individuals and groups experience and use design products and services. There is also an interest in better understanding what supports a given behavior among individuals and groups, what changes a behavior among individuals, and how designers behave and communicate within a design process.

This book is about design and design thinking from different perspectives. Design isn't the only term that carries different connotations—so does the term *thinking*. We think of something, we plan what to do by thinking, our thoughts formulate how to use artifacts and behave based on instructions, we think about how to formulate ourselves while we are talking. Thinking is an activity. Thinking contributes to the understanding of the environment and how we act within society. How we think is often betrayed by how we talk and act. It is not necessarily expressed explicitly, it can also be embedded in a system or culture. Thinking is not only individual, but it also exists on a societal level. This thinking is expressed through approaches, attitudes, expectations, and norms.

Design thinking comprises several ways of acting. In this book, authors with different design backgrounds within research, teaching, and practice present their views and experiences of what design thinking means. This book discusses design thinking in relation to theory, methods, and research. The three aspects are interwoven in most of the chapters, highlighting the complexity of design and the need to take several aspects into consideration while discussing it. It does not matter if it is on a system level or about single artifacts.

Considering that design embraces such broad areas, in order to come up with design solutions that fulfill needs and requirements, design needs to involve different aspects and competences. The involvement of the intended users of a product or service has been discussed based on examples of where users have been involved and how they have contributed to the solution. The first part of the book problematizes Design Thinking from various perspectives. It begins with a chapter by Petra Badke-Schaub where she analyzes the concept of Design Thinking in relation to how it can support the designer in the process of designing a successful product or service. Since there are no pre-existing validations of this assumption, she recommends a human thinking process that provides a scientific basis for how designers think. This comes with the assumption that designers' thinking is a part of regular thinking and, consequently, other cognitive aspects become relevant.

In the second chapter, Gabriela Goldschmidt discusses design thinking in combination with critical design. She elucidates the differences between design thinking as a way of thinking among designers, and design thinking as a method that is used in business and education in order to maximize innovation. She also argues for the need of an evidence-based understanding of the cognitive aspects of designers that is based on the documentation of a design process. By using linkography, it is possible to gain a deeper understanding of the design process and how designers think and work and come up with new products. In the following chapter, Andrew Dillon and Marian Sweeney argue that design thinking outcomes often result in insufficient analysis and synthesis since design thinking not only lacks a knowledge-based research method but also a theoretical foundation within user psychology. They suggest that, in order to maximize the value of design thinking as a problem-solving approach, it is important to leverage inductive and deductive research methodologies during appropriate phases in the process. In another chapter Åsa Wikberg Nilsson argues that, since design affects people in different aspects of their daily lives, design thinking needs to be explored from a critical perspective to be able to deconstruct what norms are embedded in design and its results. Drawing from theories on social norms, privilege, and oppression, she discusses how situatedness is a consciousness about the meaning and representation in relation to stakeholders. In the fifth chapter Yvonne Eriksson and Anna-Lena Carlsson discuss information design in relation to communication and situated awareness. The authors discuss the relationship between communication, information, and challenges in creating meaningful information which has an effect on people and the possibility to create situated awareness. In line with Dillon and Sweeney, Eriksson and Carlsson argue that several methods and theories are needed in information design since design thinking does not offer methods for evaluating design.

Åsa Wikberg Nilsson and Yvonne Eriksson develop the discussion about norms further in a specific chapter about the influence of gender in relation to design thinking and how norms are embedded not only in products, but also in representations of them, such as in product or service advertisements. In the following chapter Ulrika Florin advocates for design dialogues during design processes in order to take a holistic perspective into consideration, and not just focus on single challenges or problems when designing products or services. She demonstrates how the method could be useful for sustaining anchored designs that promote a future democratic society, built upon a combination of design thinking and systems thinking. A system thinking perspective is also present in the chapter by Christine Philips, where she argues that operational research would benefit from the use of design and design thinking in research and practice. Design and design thinking in operational research can offer a potential improvement when it comes to the integration of operational research with society as a whole.

Two of the contributions in this anthology discuss welfare technology and usability, as well as the understanding of technology among elderly people. Riika Saurio, Lea Hennala, Satu Pekkarinen, and Helinä Melkas highlight the need for information design in order to implement usable welfare technology. Whereas Marie Sjölinder discusses how a design thinking approach will bring the users' needs into the design process, as well as a joyful experience.

In the last chapter, authors Anne Wallisch and Kristin Paetzold describe how design thinking in engineering education creates awareness among students on how product development benefits from a design thinking perspective. There are some overlaps between the chapters, where authors describe the same terms and theories but in slightly different ways. This gives the reader a good insight into the fact that the terms are not fixed in meaning and, depending on which discipline one belongs to, their usage can differ. This book does not include cross-references, and every single chapter may be read independently.

CHAPTER-1
Design Thinking between Myth and Methodology

Petra Badke-Schaub

Design Thinking and design research

During the last century designing has gone through major changes, from a predominantly individual creative process to a multidisciplinary interaction process, working across different disciplines such as engineering, economics, and social sciences, but also across IT, education, and medicine.

In the late 80ies Tim Brown, the father of Design Thinking, stated in his book "Change by Design" (2009, 2017) that the problems we currently encounter in society, industry and the environment are too complex that they can be solved with the same methods we used in times gone by. This is even more true when looking into the future of mankind. Brown (2008, 2009) describes the need for change in design as the main challenge of the twenty-first century. This change should be a shift from the hierarchical and cost-efficient nature of Design Thinking to the concepts of flexibility and adaptability—which are much closer to the needs of business and management in the complex world. This statement is convincing for the designer but may not be convincing for the manager: as long as design research has not shown that the impact of Design Thinking finally leads to the desired innovation and the expected business success.

Key propositions of Design Thinking

Although there exists several different models of design thinking (see for example Efeoglu et al., 2013) here the analysis and discussion are focused on Brown (2008, 2009, 2017). Brown defines Design Thinking as "a discipline

---

Professor for Design Theory and Methodology, TU Delft - Faculty Industrial Design Engineering, Department Design Organisation Strategy, Landbergstraat 15, 2628 CE Delft.

that uses the designer's sensibility and methods to match people's needs with what is technologically feasible and what a viable business strategy can convert into customer value and market opportunity" (Brown, 2008). Brown stresses four basic premises of the Design Thinking approach:

A. Design thinking is equally relevant for designing products and spaces, as to the design of systems or dealing with more abstract problems such as services. Coping with systems and services, both aspects are now gaining more relevance for the customer and thus for the designer.
B. Design Thinking is disruptive innovation to gain a competitive advantage on the global market. This statement has been claimed already decades ago and as such it does not provide new insights, nor does it point to new behavioural strategies. In fact, it has already been more than forty years when practitioners in engineering design developed the first methodologies, that aimed at supporting the design process and, consequently, the development of innovative products. Furthermore, at the same time, 1952, Alex F. Osborn, the godfather of brainstorming, published the book "Wake up your mind: 101 ways to develop creativeness." Osborn defined three essential steps to reach a creative outcome and to arrive at a successful result: fact finding, idea finding, solution finding.
C. Design Thinking is human-/user-centred. "Design thinking is valuable not just in so-called creative industries or for people tasked with designing products. Rather, it is often a powerful tool for customers to save more, or developing a compelling narrative for public-service campaigns when applied to abstract, multifaceted problems: improving a guest experience at a hotel, encouraging bank" (Brown, 2009).

The last premise refers to the most relevant proposition of the Design Thinking approach. The focus is on the user whilst the designer seems to play the second violin. There is an assumption that designers have a kind of innate competency across different disciplines such as managerial tasks but can be done by different people other than designers. Hence Design Thinking is an activity reserved for the designer but can be, or better needs to be, done also by other people involved in the innovation of products and services development processes.

Design Thinking as thinking process

Obviously Design Thinking is an activity that cannot be observed directly but it is more than pure thinking. It needs a framework which allows to understand why certain activities occur—based on this knowledge they can be observed and interpreted.

Obviously, it is not easy to investigate such a complex and dazzling concept in a scientific way. There are almost no studies so far which have made successful forecasts of important economic developments. The same problem occurs when the impact of the use of Design Thinking should be forecasted. The expectations would be that a more creative and improved outcome (in any sense) could be assessed. These assumptions are not precise enough to end up with a clear result.

However, especially in complex situations a defined forecast would be helpful to find out what the results of the specific application are. If there is an attempt to investigate these complex, general, and partly very abstract propositions, an operationalisation of Design Thinking is necessary, which would make it applicable for thinking in if…then scenarios which usually lead to unexpected results.

Thus, the question arises: if it is not possible to formulate rules for specific situations, how can then Design Thinking be or become a (design) method? Or do we only believe in the benefits of its application? Then, the method is a myth (Norman, 2010). Against this background, Norman asks about the value of Design Thinking. Norman states that the main purpose of Design Thinking is a public relations term for good old-fashioned creative thinking. Thus, is Design Thinking a method, a strategy, a vision, an educational program, or something else?

In the remainder of this contribution an empirical study (Cardoso et al., 2014) presents components of thinking that link design research with Design Thinking and design education. This study reveals that the concept of design thinking is valid for each of the context variables but needs stronger empirical studies to understand and connect these components and to enlighten the differences in context and their implications on the outcome.

Design Thinking is question asking

The reported study here was intended to investigate design thinking as an enquiry driven process (Eris, 2004). The setting was a multidisciplinary group of seven students, tasked with designing new concepts for birthday celebrations. Questions formulated during designing are related to cognitive processes of problems solving, approaching ideas creatively and making decisions about possible design directions but also about the learned procedures on how to tackle the problem. The type of questions was assessed and categorised according to existing taxonomies when searching for and analysing questions that emerged in the existing dataset. The dataset analysed were eight sessions of 40 minutes each and one session of ten minutes of a five-day project. Finally, three types of questions could be distinguished.

The questions can be categorized in a framework that ranges from varying types of low to high-level questions. Consequently, this research aims to understand and characterise the idea generation phase of designing through an inquiry-based framework, and thus explores how inquiry might facilitate problem framing, reframing, and idea generation in such settings.

Low-level Questions (LLQ) and Direct Reasoning Questions (DRQ): The answer is known in both types of questions, which is their commonality not necessarily based on the subject of the question, but by something else. Such questions are characteristic of convergent thinking, where the questioner is attempting to converge on the facts. The answers are expected to hold truth-value since the questioner expects the answering person to believe his/her answers to be true. DRQs are different from low-level questions in the sense that the latter are used to communicate and confirm what is known, whereas the former are used to provide causal explanations of facts. For instance, "Why can't this material be exchanged?" is a DRQ. However, questions that are raised in design situations can operate under the converse premise: that, for any given question, there exist, regardless of being true or false, multiple alternative known answers *as well as* multiple unknown possible answers that are yet to be created. The questioner's intention is to disclose the alternative known answers, and to generate the unknown possible ones. Such questions are characteristic of divergent thinking, where the questioner attempts to move away from the facts to the possibilities that can be generated from them. Eris (2004) termed these types of questions generative design questions (GDQs), for instance, "How can we find out whether we can use this material?" is a GDQ (*method generation* category).

The analysis of the results reveals that high level questions trigger specific turning points during problem (re)framing. Prior to the acceleration of those turning points, the group seemed to arrive at a level of reflection that dissatisfaction with the current status steered their input, a situation that facilitates the formulation of high-level questions. As result a simple model can be developed showing how to capture this interplay between reflection, complexity, satisfaction, and high-level questioning.

Conclusions

Is Design Thinking the same process as the 'normal' thinking process? And can design methodology use the same information on how to support the designer?

Yes and no: yes, because the basic elements of human thinking are the same providing insights from psychological assumptions of how designers think. Thus, the designers' thinking process needs to be analysed in much more detail, and for many different categories of situations.

No, because Design thinking is too much focusing on general assumptions, but there are more questions to be asked (and answered) such as: Which other cognitive, motivational, and emotional aspects may be relevant for the designers' thinking? As an example discussed before the analysis of question-asking in design allows deeper insights into the way of thinking and allows hypotheses how Design Thinking can be steered by different design methods (Cardoso et al., 2016).

Design Thinking research should bring to life a master research program that determines the concepts of flexibility and adaptability disclosing the impact of different factors on the design process and product. As design methodology has to cope with vague concepts in complex field it will deliver only a small contribution to a better understanding of Design Thinking. If Design Thinking cannot be operationalised, the impact of this concept can hardly be tested in a scientific way. Aiming to support the designer design methodology needs information about the influencing factors and their impact on the design process and the design output. From these empirical results design research can define where and when and to which degree certain elements of flexibility are recommendable and when flexibility is a restless shifting from one situation to another. The integration of many interrelated variables might deliver a better picture of the situation at hand and allows answers to if…then situations for the future.

References

Bhatnagar, T. and Badke-Schaub, P. 2017. Design thinking and creative problem solving. undergraduate engineering education in India: The need and relevance. International Conference on Research into Design, pp. 953–967.
Brown, T. 2008. Design thinking. Harvard Business Review, 86(6): 84–92.
Brown, T. 2009. Change by Design: How Design Thinking Transforms Organizations and Inspires Innovation. HarperCollins.
Brown, T. 2017, 2nd. Change by Design: How Design Thinking Transforms Organizations and Inspires Innovation. HarperCollins.
Cardoso, C., Badke-Schaub, P. and Eris, O. 2016. Inflection moments in design discourse: How questions drive problem framing during idea generation. Design Studies, 46: 59–78.
Efeoglu, A., Moller, C., Serie, M. and Boer, H. 2013. Design Thinking: Characteristics and Promises.
Eris, O. 2004. Effective Inquiry for Innovative Engineering Design. Boston, MA: Kluwer Academic Publishers.
Flammer, A. 1981. Towards a theory of question asking. Physiological Research, 43: 407–420.
Martin, R.L. 2009. The Design of Business: Why Design Thinking is the Next Competitive Advantage. Harvard Business School, Cambridge, MA.
Osborn, A.F. 1952. Wake up your mind: 101 ways to develop creativeness.

Chapter-2
Critical Design and Design Thinking vs. critical design and design thinking

Gabriela Goldschmidt

Introduction

Design Thinking and Critical Thinking are both terms that are used as titles for methods or approaches meant to promote innovation in the design of new products, systems or services. Design Thinking is an orderly, teachable methodology (of which there are various varieties) that is meant to be applicable to any design process; in fact, it is tailored to be applicable to solution development in general in the business world (Leifer and Meinel, 2018) and even in education (Royalty et al., 2020). Critical Design relates to cases of extreme innovation, wherein the aim is to propose a radical, awareness-changing concept without necessarily worrying about its commercial viability. Unlike Design Thinking, which is a practical guide to more innovative performance mostly in business, Critical Design is embedded in design and artistic practice and its radicality is meant as a provocation, to help the viewer think outside the box and envision new conceptualizations of design challenges. In both cases the focus is on action to be taken.

We learn from history that designers in many disciplines do not take well to prescriptive methods and tend to either ignore them or bypass them; they take shortcuts based on their experience and rely on intuition where possible. Moreover, there is no conclusive evidence that the effort involved in using methods pays off by leading to superior outcomes. Nor can we ascertain that pieces that were recognized as landmark Critical Designs have an impact other than the buzz they create.

Technion – Israel Institute of Technology, Email: gabig@technion.ac.il

In this chapter we propose to shift attention from prescriptive methods to diagnostic observations. We shift from an attempt to prescribe what designers should do, or use as inspirational models, to an effort to understand what designers actually do on the fly, especially from the cognitive point of view, and what in design processes has the most impact on the outcome. Thus, we return to the original meaning of design thinking, i.e., a look at thinking processes of designers. In a similar manner we refer to critical design as the disclosure of what within a design process is critical to its outcomes, in the sense that it matters most. We use lower case letters for design thinking and critical design in their original connotation, to contrast with uppercase letters for the borrowed meanings in the case of Design Thinking and Critical Design as specific approaches or methods.

We start by describing and instantiating Critical Design and critical design, and then move to Design Thinking and design thinking. A case study is presented next, which explains critical design as relying on critical moves and as a derivative we also introduce the concept of critical path. This becomes possible using a linkographic analysis, which is explained further in the text. We conclude with some thoughts on design thinking, critical design, and innovation.

What is C/critical D/design?

Critical Design is a term coined by Anthony Dunne in his book 'Hertzian tales: Electronic products, aesthetic experience, and critical design', first published in 1999 by the Royal College of Art and later as an MIT Press book (2005). The term was later picked up by others, but its adoption is limited. Critical Design is hard to define: In their oft-cited manifesto of 2007 Dunne and Raby say, "it is more of an attitude than anything else, a position rather than a method", a stance that is also maintained by Jacobsone (2017, 2019). However, Critical Design had also been defined as "research through design methodology" by Bardzell and Bardzell (2013). Dunne and Rabby (2007) explain Critical Design by saying what it is not: "Its opposite is affirmative design: design that reinforces the status quo." Critical Design, then, is design that challenges the status quo, defies narrow assumptions, values, and preconceptions; it aspires to innovate by proposing designs that are critical of prevailing social norms, disregard current technological limitations, and offer a fresh look on what is possible or acceptable. The goal is "mainly to make us think. But also raising awareness, exposing assumptions, provoking action, sparking debate…" (Dunne and Raby, 2007).

Critical Design did not spring out of nowhere. It was preceded by Radical Design, a movement that was active in Italy in the 1960s and 1970s and was highly influential, especially in the realm of furniture design (Didero, 2017). It included designers, architects and artists who redefined everyday environments and introduced new concepts of form, materials, color and so on. Radical Design, like Critical Design later, was speculative, futuristic, Avant guard and often conceptual more than useful. It yielded mostly exhibition pieces rather than commercial products; however, a group like Memphis, that sprang out of Radical Design (Milan, 1980s), was also very active commercially. Its innovative and provocative designs for products, interior spaces and buildings evoked intense discussion and continue to be influential even today (Bingham, 2019).

Although Dunne and Raby claim that Critical Design is not art, there is no denying that the border line between the two is rather fuzzy and so called 'experimental design' can sometimes be interpreted as either critical design or critical art. In either case the point is to show different thought directions, to expose new options and alternative lifestyles at all levels, and to that end many such design projects are rather provocative. By way of example, let us consider work by Natalie Jeremijenko, who defines herself (among others) as an experimental designer. Jeremijenko's background and education is in the exact sciences and engineering but she is best known as an experimental artist. The work we present here is called 'Tree logic' (Figure 1). It is an installation in front of the Mass MoCA—Museum of Contemporary Art, in North Adams, Massachusetts, since 1999.

The project consists of maple trees planted in large metal containers and suspended, inverted, from a truss that was built for this purpose. Water and nutrition are provided through an elevated tube system. The trees obey the laws of nature and their branches grow upward toward the sun, creating an unfamiliar and non-iconic representation of trees

Figure 1. Natalie Jeremijenko, 1999. 'Tree Logic', suspended inverted trees, MASS MoCA, North Adams, Massachusetts (Photo: G. Goldschmidt, 2004).

(the trees are replaced from time to time, when their roots overgrow the capacity of the containers). Jeremijenko calls our attention to the dynamic interplay between gravity and the phototropic imperative, which normally go hand in hand, but here they are in competition and must overcome an impediment: we are reminded to not take anything for granted.

Bardzell and Bardzell (2013) suggest that Critical Design takes after other 'critical' movements, in particular that of Critical Theory, starting with the Frankfurt School, which comprised philosophers and social theorists who championed the Marxist tradition in Western Europe starting in the 1920s (and later in other parts of the world). Their social theories were critical, according to the members of the movement, insofar as they sought to liberate and emancipate human beings and transform social conditions in all walks of life (Stanford Encyclopedia of Philosophy, 2005). Critical Design adopted the wish to change, to question all prior norms and assumptions, and to make people re-think about the world and artefacts around us.

Critical design, without uppercase letters, is something entirely different. To define it, I would like to use the word 'design' as a verb: to design; then we can ask what is critical designing? Critical design(ing) is that which, in the process of design, is critical to an effective completion of the process, yielding a successful outcome. Behind this definition lies the assumption that during the process, the designer is engaged in a large number of design acts, some of which more important than others. In the front edge of the process, which is the phase we would like to focus on, the design process is a search for options, ideas, relevant information and precedents, alternative solution options and ways to realize them. Since design problems are ill defined, the process cannot be linear and oftentimes the designer cannot know in advance whether any particular path will lead to a satisficing solution, and therefore a lot of exploration and experimentation is involved. Usually, it is only after the fact that we can point to those acts, from amongst the large number of all design acts, that were decisive in leading to an optimal solution. The acts in question may be of varied types: questions, speculations, proposals, decisions, and more.

Once we pinpoint the critical design acts (depending on the scale of the investigation, they can vary in duration from single moves of a few seconds at the cognitive scale to entire episodes in longitudinal examinations), we can string them together into a critical path. Critical path here means something different than what the same term denotes in project management and operations research. In those fields, critical path is a method, a "technique for analyzing, planning, and scheduling large, complex projects" (Levy et al., 1963). It is used primarily to schedule jobs in a project in the most efficient way in terms of cost and time savings. In the critical path method, the critical components are established ahead of time.

In our use of the term, they can be identified only in retrospect, based on documentation of the project's chronology.

In our terminology, even more important than a critical path are the components of which it is composed, namely, acts of design. Since we are interested in the cognitive scale, the acts in the process of design are moves (as in chess) which, as mentioned above, are very short. We call them design moves; their duration is usually less than 10 seconds (Goldschmidt, 2014). Those moves which in retrospect are the most decisive to the outcome are critical moves. How are they determined? One simple way is to get well acquainted with the process, which is usually captured in a recorded protocol of all verbalizations in a given episode, and then to subjectively judge which were the most important moves that led to the solution the designers have reached. The judgment is based on a very good acquaintance with the process and the outcome, and expertise in the discipline in question. We use a different approach to the establishment of critical moves, based on the network of links among all moves, on which we elaborate further on in this text.

What is D/design T/thinking?

Design thinking is not a new term. However, in the past two decades it has become a household term (or dare we say buzzword?), in good currency among designers and business innovation promoters around the globe. In its current incarnation Design Thinking denotes a method meant for general use in the process of devising novel solutions for products, spaces, services (including 'experiences') or systems. More than anything, it is a business strategy (Lockwood, 2009; Martin, 2009; Verganti, 2009). In California's Silicon Valley, Design Thinking is even referred to as a movement: *The Design Thinking Movement*. It pertains to a method that is used and taught, and one is perforce reminded of *The Design Methods Movement* in Britain in the 1960s (Langrish, 2016). The raison d'être of these 'movements' was then and is now to improve design processes such that the outcome be the best possible, and in the case of the current Design Thinking Movement, the expectation is above all for an innovative outcome. However, the scope and methodology differ considerably.

The Design Methods Movement sought the 'scientification' of design by developing prescriptive optimization models, mostly aimed at achieving better products as a result of efficient methods. By comparison with the current Design Thinking Movement, the Design Methods Movement was not as business oriented, and not so focused on innovation. Today's Design Thinking is motivated by an innovativeness imperative. Both designers and managers use the term, and the methodology, of Design Thinking. There are similarities and differences between the way the design and the business communities relate to Design Thinking. In a study comparing the attitudes of the two communities, Marnina Herrmann (Herrmann and Goldschmidt, 2013) found that both communities emphasize iterative processes, collaboration, speed of concept modeling and testing through prototyping, and interaction with users. Designers see Design Thinking as a learning process; the business community sees it as a knowledge-based process. Design texts emphasize experience, exploration, experimentation, play, and even failure when it leads to learning. Designers tend to search out validity, whereas the business community seeks reliability. Business texts on Design Thinking stress innovation, while design texts center more on creativity. There is also a notable difference in the preferred mode of communication, which is mostly visual in design and verbal in business.

Design Thinking methods differ from the methods of the 1960s, though. They are less specific and more general and flexible; they are more like guidelines than detailed operational steps. Therefore, they induce less resistance, but opposition does exist all the same, and some experts predict that Design Thinking as a method will soon face the same fate as the design methods of the 1960s (Nussbaum, 2011). Design Thinking does not contribute to our understanding of the underlying cognitive processes related to designers' activities, which is what the term design thinking originally meant, and still means. Design Thinking does not explain what the particular knowledge and skills that designers possess are (Cross, 2006; Visser, 2009), how a synthesis of "new ideas from seemingly disparate fragments" (Brown, 2009, dust cover) occurs, or how designers convert ideas generated by a team into implemented solutions. Surely, even the most sophisticated observation methodology cannot answer these questions, at least not at the cognitive level. It is here that design thinking research becomes relevant, and in particular that branch of research which involves cognitive (and as of late also neurocognitive) psychology as a way to access designers' thinking, at both the individual and team settings.

Design thinking in the simple sense of how designers actually think on the fly started attracting attention because of the failure of the Design Methods Movement, and because the discipline of cognitive psychology had come of age at about the same time, which pointed at a new path to understanding designing. The acknowledgement of the failure of the design methods of the 1960s stemmed from the realization that such methods go against the grain of natural thinking in design. The recent incarnation of design methods in the form of Design Thinking was therefore somewhat risky to begin with, as it was not clear whether it resolves the incompatibilities between the way designers think and the use of prescriptive methods.

Following the disillusionment from the design methods of the 1960s, it became clear that it is both necessary and possible to study thinking, including in design, whereas previously this kind of thinking was considered 'magic' and impossible to fathom. Bryan Lawson published 'How designers think: The design process demystified' in 1980, and 'Psychology of architectural design' was published by Ömer Akin in 1986. Akin was the first researcher who applied protocol analysis widely in design research (following earlier smaller scale studies by Charles Eastman in the 1970s). Indeed, protocol analysis became a major research methodology in the research of design thinking. It is still used abundantly today, as it is well suited to cognitive studies and generally to behavioral studies. It received its final 'stamp of approval' with the publication of 'Analyzing design activity', an edited volume by Cross, Christiaans and Dorst in 1996, dedicated entirely to the use of protocol analysis in design thinking research. In recent years neurocognitive studies are also being pioneered in the field of design, but those are still scarce and require expertise that is not needed for protocol analysis studies.

Critical moves, critical path in linkography

Linkography is a method of notation and analysis that was developed for design processes but is applicable also for other processes (e.g., in Goldschmidt, 1990, 2014), which has become a prominent research tool in design thinking research. It is based on the notion that in its early ideation stage the design process is a search aimed at arriving at a synthesis. A synthesis is built up step by step, while exploring possibilities and constraints and assessing them. The final outcome must perforce be reached by linking different aspects together into a unified whole. Therefore, we deal with links within the design process. Linkography uses protocols parsed into bigger or smaller segments which are design moves, or other entities such as decisions, or ideas, as the units of analysis. Every pair of moves (or other units) is assessed in order to answer the question: is there a link between the two moves? A link is determined based on the contents of the moves, and common sense, resting on good familiarity with the discipline and the specific process in question. When moves are checked against previous moves, *backlinks* are detected. After the fact, we can also talk about *forelinks*, which are (virtual) links between moves and posterior moves. The total number of backlinks and forelinks in a given linkograph is equal, of course: each link is a backlink of the later move in the paired couple, and a forelink of the earlier move in the same couple of moves.

The resultant network of links among moves, which is notated in a likograph, makes it possible to identify and measure various parameters of the process which are related to its structure, effectiveness and creativity. A major feature of linkography is the identification of Critical Moves (CMs), which are moves with a large number of links. We distinguish between CMs due to backlinks (<CM) and CMs due to their forelinks (CM>). We establish a threshold level of the number of links (backward or forward) that qualify a move as critical; the higher the threshold, the smaller the number of critical moves at that level (for level x, the notation is CM^x). Critical moves are seen as the most important moves in a sequence, because if links are important, then moves that generate the largest number of links, thereby contributing significantly to the achievement of synthesis, are the most important moves in that sequence.

The chain of CMs in a given process (or portion thereof) is referred to as a critical path (see previous section).

Figure 2 is an illustration of a short linkograph. The Critical Path at the level of CM^3 consists of moves 3, 7, 11, 12. Move 3 has 6 forelinks (and 1 backlink) and can be described as $CM^6>$. Move 7 has 5 backlinks (and 2 forelinks) and is therefore a $<CM^5$. Moves 11 and 12 have only 3 links each, both backwards, and are therefore $<CM^3$s.

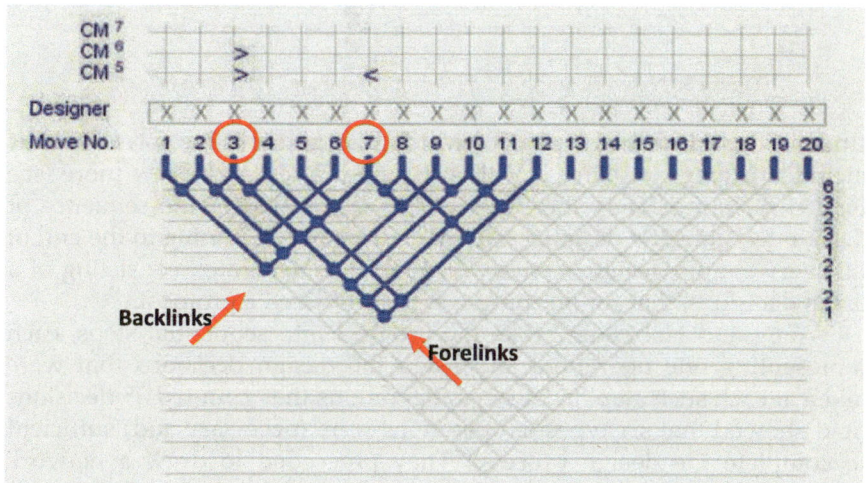

Figure 2. Demo linkograph highlighting critical moves 3 and 7, qualified as at least CM^5s.

Case study: Wang and Habraken's critical path

John Habraken and his PhD student at the time Ming-Hung Wang, set out to study the basic operations that they believed are sufficient to carry out designing, based on decisions that are being made during the design process (Wang and Habraken, 1982). They suggested six such operations (the operations are of no interest to our current topic and therefore we do not elaborate on them here). In a think-aloud study that was video-taped they asked a designer to arrange nine given form-variants (furniture or defined space with dedicated furniture) in a living—dining space in a

14 *Different Perspectives in Design Thinking*

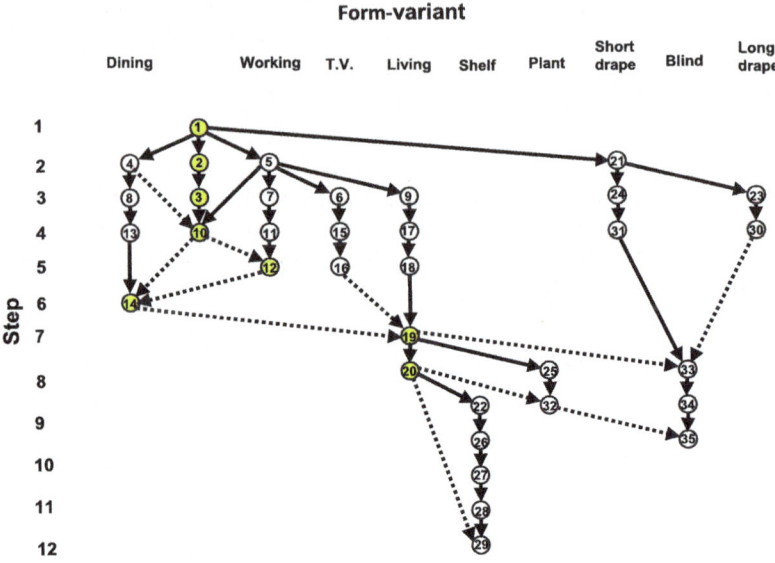

Figure 3. Network graph, adapted from Wang and Habraken (1982).

student's apartment, the plan of which was given; these form-variants included dining, working, T.V., blind, long drapes, and a few more (see Figure 3). Immediately thereafter a list was created of the sequence of design decisions the designer had made from the beginning to the end of this very simple design exercise. A report was generated consisting of a list of decisions, that for our purposes is considered a protocol.

Wang and Habraken broke the protocol into sequential steps, each representing one operation, and listed the design decisions that were taken at each such step. In 24 steps/operations they counted 35 decisions (and showed that six types of operations were (necessary and) sufficient to complete the design process). They proceeded to draw a network graph that represented the process according to their analysis, wherein nodes stand for decisions and connecting arrows stand for operations. Next the operations that were carried out more or less at the same time were consolidated into one step (i.e., the designer attended to more than one form-variant almost simultaneously. Previously they were listed separately). As a result, the number of steps was reduced to 12, in which the 35 decisions were taken. Figure 3 shows the final network graph.

At each step, the designer made decisions (represented by nodes) concerning one or more form-variants. To reach decisions the designer carried out operations (represented by arrows). Only solid arrows represent coded operations (which refer to either 'sequential relation' or 'logical precedence'). Dotted arrows represent 'dummies' which refer to "information flow between nodes, without time and labor costs." Since

Wang and Habraken (1982) stress the importance of these dummies, we treat them as full-fledged equivalents of coded operations.

Wang and Habraken proceeded to determine a *critical path* that includes eight major decisions taken along the process: 1 – 2 – 3 – 10 – 12 – 14 – 19 – 20. This critical path is based on their own expert assessment of the process. The decisions that were included are identified by their sequential numbers (we leave out the decision contents, which are not relevant to this discussion).

The information in Figure 3 served us to construct a linkograph that is shown in Figure 4. However, in a linkograph nodes are links, whereas Wang and Habraken's nodes—design decisions—are constants in the linkograph. Decisions here are equated with moves and Wang and Habraken's operations are links (both solid and dashed lines). Figure 4 depicts the resultant linkograph.

Based on the linkograph we can single out those decisions that have the highest number of links. I borrow Wang and Habraken's use of the term critical path but establish an independent critical path; this critical path is not based on an educated assessment of the process as in Wang and Habraken, but on a count of decisions with at least three (a threshold that was established for this case) backlinks or forelinks.

In the linkograph the nodes represent links among the design decisions in Wang and Habraken's study. The diagonal lines connecting decisions with links are gridlines that have no operational significance (as opposed to Wang and Habraken's network graph); they facilitate the count of links associated with each decision, and they help perceive the structure of the process.

Based on the count of links we obtain the following critical path at CM^3: 1 – 5 – 10 – 14 – 19 – 20 – 33. Although the two critical paths, that of Wang and Habraken and that based on the linkograph, are not identical,

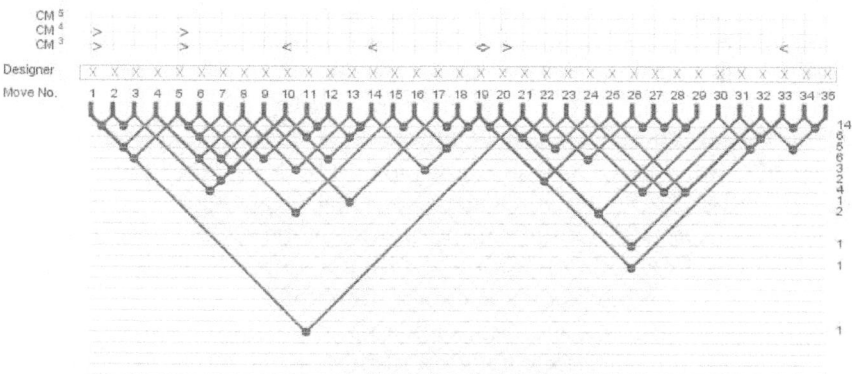

Figure 4. Linkograph based on Wang and Habraken's network graph (see Figure 3).

Table 1. Critical Paths, Wang and Habraken's expert assessment, and linkograph.

Decision no.	1	2	3	5	10	12	14	19	20	33
Critical Path Expert assessment Wang and Habraken	x	x	x		x	x	x	x	x	
Critical Path Linkograph, CM^3	x			x	x		x	x	x	x

they are reasonably similar (five identical decisions out of eight and seven in the critical paths, respectively), see Table 1. Given that decision 5 is in fact the implementation of decisions 2 and 3, which brings the two critical paths even closer, the similarity between the two critical paths reaches a high 70%.

Thus, we are able to demonstrate the close relationship between criticality of decisions as determined by the intensity of interlinking, and criticality as determined independently by other, unrelated criteria, in this case expert assessment. This result supports the notion that important instances of designing and design reasoning are indeed distinguishable by the high number of links they form, backwards or forwards, compared to other moves in the same sequence.

Thoughts about design thinking and design criticality in innovative design

The starting point of any discussion of Design Thinking—the method—is innovation. Design Thinking was formulated with the purpose of promoting innovation in all lines of business. The label is somewhat misleading and may have been chosen because designers were the prominent pioneers of this method: its origin is often attributed to IDEO, then a young design consultancy in Palo Alto. According to Tim Brown, IDEO's Executive Chair: "Design Thinking is a human-centered approach to innovation that draws from the designer's toolkit to integrate the needs of people, the possibilities of technology, and the requirements for business success." (http://designthinking.ideo.com/#design-thinking-in-context). Calling Design Thinking a movement, Barry Katz who is also associated with IDEO, adds that "one does not have to be a designer to think like one" (Katz, 2015). Thus, we clearly see that Design Thinking, which is really 'Design-*like* Thinking', can by no means be equated with design thinking tout court, which is a study of thinking processes while designing. Because the assumption is that designers' thinking is creative and they are better equipped to bring about ideas that promote innovation, businesses are interested in harnessing this kind of thinking in order to help them innovate, which is an economic imperative nowadays. A simple Google search reveals that there are hundreds of innovation consultancies around the developed world, all of which owe their existence and livelihood to the overarching sentiment of businesses that unless they innovate, they are destined to perish. So why is a method necessary, if designers know how to think "the right way" in the first place? At the same time, we do not forget that most designers are not, as we learned from the failure of the Design Methods Movements half a century ago, great fans of methods (see also Goldschmidt, 2008).

Here we should tarry for a moment to look at the different circumstances that brought about the two movements—the design methods movement and the Design Thinking (-like) movement. In the 1960s designers were conceived of as professionals who worked alone. That is, design ideas were for the most part generated by a chief designer in a design firm and were then worked out by his or her subordinates. As design tasks grew larger in scope and complexity, the observation was that designers did not have enough knowledge or experience to deal alone with such problems and therefore methods were sought. A few decades later the single designer's predominance was largely mitigated and reserved only to very special cases of star-designers; in the case of architecture, for example, they were indeed called starchitects. But in most cases the understanding that design tasks were too complex and demanding for one person to lead the way to a solution finally settled in and design is henceforth almost always delegated to teams, in many cases multidisciplinary ones. Such teams have been shown to perform more creatively than uni-discipline teams (e.g., Reiter-Palmon and Leone, 2019; Somech and Drach-Zahavy, 2013). Teams, especially if they are inter- or multidisciplinary, cannot be likened to individuals in the sense that they produce multiple mental models of the task at hand and its solution, and cannot really succeed in their mission unless they manage to generate shared mental models (Badke-Schaub et al., 2007). To produce shared mental models that guide the development of solutions to problems when a team is concerned, using effective prescriptive methods is very helpful in converging individual mental models into shared ones. From this point of view, we may propose that Design Thinking was conceived in order to facilitate the creation of shared mental models in teams composed of members with diverse professional backgrounds: the method is a convergence agent.

However, with all its advantages Design Thinking does not help us understand design thinking. While Design Thinking may aid in bringing about innovation, design thinking helps in understanding how innovative ideas are generated. Methods that served, and still serve this kind of research extremely well are protocol analysis and linkography. However both protocol analysis and linkography have the severe limitation that they are extremely labor intensive; to date they are based on human judgment of the content of segments of verbalization by designers at work. Efforts to automate protocol analysis using various AI and other digital means have hitherto been largely unsuccessful. The same applies to linkography as well. Therefore, most studies that explore design cognition are based on limited datasets, to be manageable when used with these manual methods. New AI tools are currently penetrating research of this type, allowing the analysis of large datasets. These tools are based on network theory (see Taura and Nagai, 2013), on a latent semantic approach (Dong, 2005) or on linguistic analysis involving machine learning (Mikolov et al., 2013); that is, the system is "taught" a vocabulary in a particular field and learns to distinguish the use of words or expressions in a context-sensitive manner. In the field of design this line of research is still in its infancy and the coming years will undoubtedly yield interesting results.

We remain with the question of criticality. When team discourse is analyzed, the contents are usually coded into various categories that depend on the purpose of the study. Part of the discourse reflects team social issues, and even to jokes or other discourses that humans engage in to ease the flow of work. AI tools can point to key words, frequent expressions, etc., but determining what in the discourse is critical in the sense we outlined above, namely, what is indispensable to the success of the outcome and its level of creativity, is not an easy task for currently available technologies. We can safely assume that eventually this will be achievable. Linkography offers a unique way to determine criticality of segments of the design process, based on moves that are especially effective in inducing synthesis. If linkography is automated, and therefore useful for much larger datasets than is currently the case, the concept of critical moves will be easy to work out. We will then have at our disposal the new kind of critical paths, those created post-factum to shed light on the nascent moves that eventually contributed to innovative concepts. Some of them may be reoccurring and reusable, maybe predicting important breakthroughs beyond the development of one specific product or one system or one service.

To conclude, methods come and go—they are replaced when better and more potent ones are developed. With the rapid pace of technological evolution, the lifecycles of methods are probably also getting shorter than in the past. The human mind does not evolve as fast of course; cognition, and in our case design cognition, is much more steadfast than are the methods we use in design or in design research. Therefore, it will continue to be important to study design thinking and its critical instances long after Design Thinking is replaced by the next best methodology. Critical Design will continue to challenge, provoke, and inspire us, but critical design will illuminate the way to innovation.

References

Akin, Ö. 1986. Psychology of Architectural Design. Pion, London.
Badke-Schaub, P., Neumann, A., Lauche, K. and Mohammed, S. 2007. Mental models in design teams: A valid approach to performance in design collaboration? CoDesign, 3(1): 5–20.
Bardzell, J. and Bardzell, S. 2013. What is "Critical" about critical design? pp. 3297–3306. *In*: Bødker, S., Brewster, S., Baudisch, P., Beaudoin-Lafon, M. and Mackay, W.E. [eds.]. Proceedings of CHI 2013: Changing Perspectives. The association for Computing Machinery, NYC.
Bingham, C. [ed.]. 2019. More is More: Memphis, Maximalism, and New Wave Design. teNeues, Kempen, Germany.
Brown, T. 2009. Change by Design: How Design Thinking Transforms Organizations and Inspires Innovation. Harper Collins, New York.
Cross, N., Christiaans, H. and Dorst, K. [eds.]. 1996. Analyzing Design Activity. Wiley, London.
Cross, N. 2006. Designerly Ways of Knowing. Springer Verlag, London.
Didero, M.C. 2017. Super Design: Italian Radical Design 1965–75. The Monacelli Press, New York.
Dong, A. 2005. The latent semantic approach to studying team communication. Design Studies, 26: 445–461.
Dunne, A. 1999/2005. Hertzian Tales: Electronic Products, Aesthetic Experience and Critical Design. RCA, London (1999); MIT Press, Cambridge, MA (2005).
Dunne, A. and Raby, F. 2007. Critical design fac. http://dunneandraby.co.uk/content/bydandr/13/0 Retrieved 3 June 2020.
Goldschmidt, G. 1990. Linkography: Assessing design productivity. pp. 291–298. *In*: Trappl, R. [ed.]. Proceedings of the Tenth European Meeting on Cybernetics and Systems Research. World Scientific, Singapore.
Goldschmidt, G. 2008. Sketching is alive and well in this digital age. pp. 29–43. *In*: Poelman, W. and Keyson, D. [eds.]. Design Processes: What Architects and Industrial Designers can Teach each other about Managing the Design Process. IOS Press, Amsterdam.
Goldschmidt, G. 2014. Linkography: Unfolding the Design Process. MIT Press, Cambridge, MA.
Herrmann, M. and Goldschmidt, G. 2013. Thinking about design thinking: A comparative study of design and business texts. pp. 29–40. *In*: Chakrabarti, A. and Prakash, R.V. [eds.]. ICoRD'13: Global Product Development. Lecture Notes in Mechanical Engineering Series. Springer Verlag, New Delhi.
Jacobsone, L. 2017. Critical design as approach to next thinking. The Design Journal, 20(sup 1): S4253–S4262.
Jacobsone, L. 2019. Critical design as a resource. Adopting the critical mind-set. The Design Journal, 22(5): 561–580.
Katz, B.M. 2015. Make it New: The History of Silicon Valley Design. MIT Press, Cambridge, MA.
Langrish, J.Z. 2016. The design methods movement: from optimism to darwinism. pp. 51–64. *In*: Lloyd, P. and Bohemia, E. [eds.]. Proceedings of DRS2016: Design + Research + Society: Future-Focused Thinking vol. 8. Design Research Society, London.
Lawson, B. 1980. How Designers Think: The Design Process Demystified. Butterworth, Oxford.
Leifer, L. and Meinel, C. 2018. Looking further: Design thinking beyond solutions. pp. 1–12. *In*: Meinel, C. and Leifer, L. [eds.]. Design Thinking Research. Understanding Innovation. Springer, Cham.
Levy, F.K., Thompson, G.L. and Wiest, J.D. 1963. The ABCs of the critical path method. Harvard Business Review, 41(5): 98–108.
Lockwood, T. [ed.]. 2009. Design thinking: Integrating innovation, customer experience, and brand value. Design Management Institute. Allworth Press, New York.

Martin, R.L. 2009. The Design of Business: Why Design Thinking is the Next Competitive Advantage. Harvard Business School, Cambridge, MA.

Mikolov, T., Chen, K., Corrado, G. and Dean, J. 2013. Efficient estimation of word representations in vector space. In Proceedings of International Conference on Learning Representation, arXiv preprint arXiv: 1301.3781.

Nussbaum, B. 2011. Design Thinking is a failed experiment. So what's next? FastCompany, https://www.fastcompany.com/1663558/design-thinking-is-a-failed-experiment-so-whats-next Retrieved 27 February 2021.

Reiter-Palmon, R. and Leone, S. 2019. Facilitating creativity in interdisciplinary design teams using cognitive processes: A review. pp. 385–394. *In*: Childs, P. and Cropley, D. [eds.]. Proceedings of the Institution of Mechanical Engineers, Part C: Journal of Mechanical Engineering Science, 233(2).

Royalty, A., Chen, H., Roth, B. and Sheppard, S. 2020. Reflective tools for capturing and improving design driven creative practice in educational environments. pp. 49–65. *In*: Meinel, C.L. and Leifer, L. [eds.]. Design Thinking Research. Understanding Innovation. Springer.

Somech, A. and Drach-Zahavy, A. 2013. Translating team creativity to innovation implementation: The role of team composition and climate for innovation. Journal of Management, 39(3): 684–708.

Stanford Encyclopedia of Philosophy. 2005. https://plato.stanford.edu/entries/critical-theory/Retrieved 11 June 11 2020.

Taura, T. and Nagai, Y. 2013. Concept Generation for Design Creativity: A Systemized Theory and Methodology. Springer Verlag, London.

Verganti, R. 2009. Design Driven Innovation: Changing the Rules of Competition by Radically Innovating What Things Mean. Harvard Business School, Cambridge, MA.

Visser, W. 2009. Design as one, but in different forms. Design Studies, 30(3): 187–222.

Wang, M.-H. and Habraken, N.J. 1982. Notation of the Design Process: The Six Operations. A Preliminary Investigation. MIT: Unpublished research paper.

CHAPTER-3

Adding Rigor to Advance Design Thinking

Andrew Dillon[1,*] and *Marian Sweeney*[2]

The rise and quick adoption of design thinking (DT) as a general problem-solving approach to innovation in business and technical domains has resulted in rapid growth of demand for appropriate education and training. User-experience (UX) designers have more traditionally been educated in user-centered design (UCD) methods, an approach that has significant overlap with design thinking but which some argue is distinct and insufficiently agile to be easily applied in industry.

Rather than treat the question of equivalence or distinctiveness as key, in the present chapter we argue that information design practices are necessarily fluid *in situ*. In particular, studies of design, from architecture to software development, have shown that it is not an entirely rational process (see, e.g., Lawson, 2006). Further, there is no one method that all designers follow. Many instances of good design involve a hybrid form of problem solving that might involve components of both DT and UCD, depending on numerous contextual factors.

We make the case in this chapter that the design thinking emphasis on outcomes often results in insufficient analysis and synthesis. Specifically, we show that well-established scientific principles concerning the psychology of user experience are not fully leveraged in the process, resulting in a form of praxis that is insufficiently informed by the knowledge base. Design thinking as currently taught, lacks a knowledge base around research methods but also a theoretical foundation on user psychology. Hence, we argue that to maximize the value of design thinking as problem-solving approach, it is important to leverage inductive and deductive research methodologies at appropriate phases in the process.

---

[1] School of Information, University of Texas at Austin.
[2] AT&T Design Technology, Austin, Texas.
* Corresponding author: adillon@ischool.utexas.edu

What is design thinking?

In its most basic form, design thinking represents an approach to group problem solving that is based on rapid and continual idea generation, usually in a context of divergent thinking that resists early judgement so as to encourage participation and discussion among stakeholders.

All definitions of design thinking share a set of steps or activities as follows:

- Empathy—an effort to learn about the users for whom you're designing and their context of use in the real world
- Definition—a phase of establishing user needs, clarifying the problem
- Ideation—Generating ideas for potential or innovative solutions
- Prototyping—Mocking up concepts
- Testing—Trying out the ideas on real users to gain feedback

These iterative steps are presented in the following visual model developed by the Interaction Design Foundation.

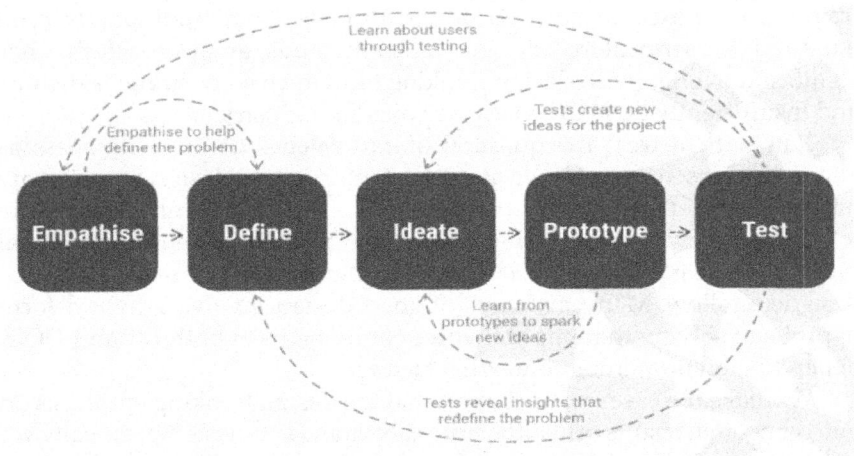

Interaction Design Foundation https://www.interaction-design.org/literature/article/stage-2-in-the-design-thinking-process-define-the-problem-and-interpret-the-results

In this flow there is inherent flexibility in the steps which ensures no two processes are the same, each step can be revisited repeatedly on the basis of outputs at other stages and it's possible that prototyping and testing can lead to complete revisions of understandings generated at the empathy stage, for example. However, the goal is to move through the process in as agile a manner as possible to deliver appropriate solutions which gain buy-in from stakeholders. We discuss each step in more detail below.

Empathy

The first phase of design thinking is empathy. The main objective is understanding needs based on user and stakeholder interviews, contextual observations and work sampling. This speaks to the involvement of an appropriate number of intended users from across a sufficiently broad range of use contexts. The work in the empathy phase demands rational and empirical research practices to ensure valid and reliable discovery of the currently unmet user needs that essentially represent the problem to be solved. In processes that begin with a design sprint or which shortchange detailed user research, the problem-to-be solved may never be properly defined. In these cases, the drivers of design solutions may not reflect the reality of users' needs in the field.

Design thinking describes resources and artefacts that support the empathy phase; tools such as empathy maps, as-is journey maps and personas may be utilized in order to embody the goals, needs, experiences, and emotions of real users for which design solutions should be considered. These tools present the 'research' in a readily consumable format for the whole team. By its democratizing nature however, DT methods do not prescribe any knowledge required for their useful application, and design teams may consist of people with little or no research background, utilizing techniques such as the 5-Whys or Why-How laddering as research methodology. A more traditional UCD researcher would understand the problems of non-random sampling and control of bias but such aspects of data collection are rarely acknowledged or given consideration within design thinking which can obviously lead to problems of appropriate interpretation and coverage.

After the initial discovery research phase, the team usually has a kick off workshop for the project. Stakeholders and researchers often present lightening talks about the unmet needs, current workarounds and pain points for users.

Define

In the second phase of a typical DT process, the design team will begin to articulate the problem that needs to be solved. A problem or need statement will establish agreement on the gap between the current state and the desired state. Problem statements are often translated, using the terminology of design thinking into 'How Might We?' statements or 'Who, What, Wow' statements, for example: "A call center representative (*the Who*) needs a way to retrieve all customer accounts (*the What*) immediately (*the Wow*) after authenticating". Once a set of How-Might-We or Need statements are accepted as encapsulating the problem, the design thinking team generate as many ideas as possible to tackle the issue outlined.

The problem definition activity is arguably the most crucial part of the design process as it identifies where all subsequent effort will be devoted. This step aims to render the problem statement more specific and actionable. Missing the right target at this point is hugely problematic because any solutions that are created will likely not address the real issue and waste resources in the pursuit of inappropriate design goals. There is little scaffolding within the design thinking framework for honing the most appropriate problem statement, other than common sense, which is generally a poor guide to understanding people's needs. The DT process however encourages swift resolution, leveraging our human nature to solve a problem as soon as we can. A concern that is often raised by critics is that most people will start 'solutioning', that is presenting possible design outcomes, even as the problem is still being articulated (Sproull, 2018).

Ideate

The generation of concepts, ideas and attributes for the design solution characterizes what is termed the Ideate phase. Here, participants are encouraged to engage in divergent thinking to generate ideas, even wild ones embracing 'magical thinking' are welcomed. The emphasis in this phase is on abundance rather than completeness or feasibility of ideas in order to facilitate imaginative solutions. Typically, the entire DT team will outline concepts on Post-it notes in a rough picture format intended to avoid rationalizing or detailing every idea. This unleashing of creativity through divergent thinking is one of the key components of design thinking. The intentional open mindedness and risk-free environment for idea submission is intended to encourage the team to consider transformative rather than incremental design.

This ideation activity is usually time-bound within a workshop setting so that evaluating the generated ideas is enabled to refine the thinking thus far. As participants describe and cluster ideas in themes or patterns of similarity, the team members try to improve the ideas by engaging in what are termed "YES, AND" statements. For example, 'YES, a self-driving hovercraft will float over rough terrain, AND if you add 360° sonar, it will also be able to avoid obstacles. Such comments are designed to accentuate the value of one idea and help tweak it in a refined direction.

Once ideas are reviewed and tweaked, the team enters a more convergent form of thinking to help focus attention on the best ideas. Generally, a voting process begins in order to galvanize consensus around which ideas seem to best solve the problem statement(s). The practicality of a physical workshop environment, with participants huddled around a wall of Post-it notes, can make it difficult to hear the descriptions or make sense of another person's rapidly sketched concept. In this active and

Adding Rigor to Advance Design Thinking

interactive milieu, convergence is generated usually by 'voting', whereby participants place dots on the ideas that seems to best fit the need.

The design thinking framework itself does not present a structured or rational methodology for assessing candidate ideas. Few if any criteria are established and voting can become somewhat arbitrary and chaotic.

The IDEO playbook on prioritizing ideas presents two 'methods' for evaluation. The first is to position all ideas on a Value versus Complexity matrix so see what emerges as low-hanging fruit (highest value for lowest

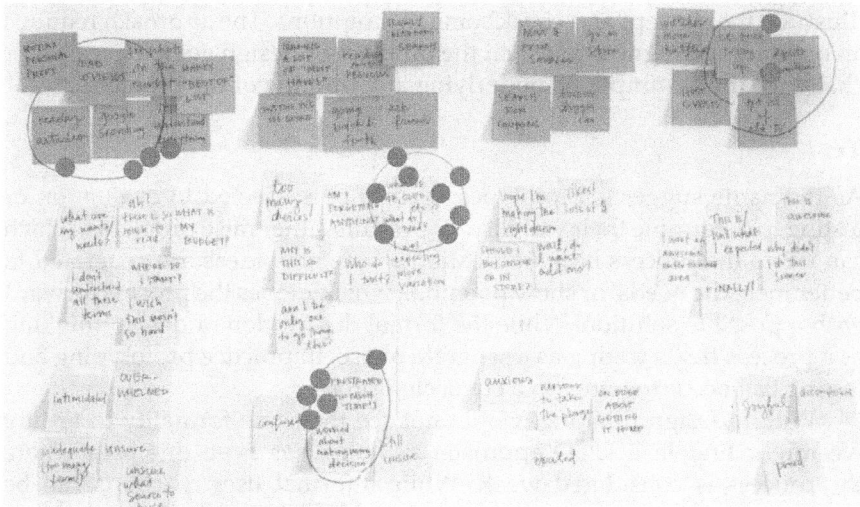

From https://www.nngroup.com/articles/dot-voting/

complexity). The second is to choose ideas that the team are most excited about. This is another example of a limitation within design thinking as preference and excitement not formally methodological, leaving considerable room for group-think and short-term consideration to determine key design choices.

Prototype

In the full design thinking process, once candidate concepts are selected the process of rapid prototyping begins to bring ideas to life. The aim is to give substance to the ideas that are deemed to have potential and to reveal how the resulting product or service might appear to users. This is a common phase for every design method although in design thinking, prototyping usually is led by the members on the team with specialized interaction design or graphic design skills. The American graphic designer Elizabeth Neumier (cited in Shapiro, 2020), emphasizes the importance of viewing design as "the process of working through a complex challenge using a succession of prototypes; i.e., *thinking by making*. This leaves a lot of room for interpretation, as it should."

In all design contexts, the iterative production of prototypes provides a way to reframe the problems and to gain an opportunity for learning and progression. Von Thienen et al. (2019) describe rapid prototyping as a means of deriving tangible results that satisfy unmet human needs. In alignment with the User Centered Design (UCD) philosophy, the design thinking approach emphasizes continual iteration of these prototyped concepts for all personas in the problem space. Prototypes help to illustrate the concept and check some assumptions. The approach requires an ongoing willingness to revisit the problem or design goal, to recast it or challenge the assumptions underlying any framing of the problem.

Test

As the name suggests, prototypes need to be subjected to evaluation or testing to determine their viability. The results either raise problems, which can move the process back to ideation where new ideas are generated to better meet the needs, or show potential, which moves the process forward with a possible solution. While the formal description of design thinking as a process treats testing as a separate phase, in practice prototyping and testing can occur together in a connected process.

Within design thinking, we do not see the same formality to testing we might find in a UCD approach and this has been one area where the process is considered weak. While informal user reactions can be extremely informative, particularly where they help generate new ideas (i.e., return the design thinking process to the ideation phase), there is rarely acknowledgement within design thinking of the value of controlled, structured evaluations such as the type of usability tests that are common in UCD.

In sum, design thinking encapsulates efforts to think freshly about problem spaces and encourages divergent thinking to generate solutions that offer significant change not simply improvements over current products or services. It is characterized by an open, non-credentialed treatment of inputs, and deferred judgement on value, which makes the process attractive to participants. It has considerable overlaps in terms of basic phases with other approaches to design but is considerably less methodological in key areas that are deemed essential by practitioners of UCD, and there are no particular knowledge or skills requirements for design thinking team members, as topic usually glossed over with vague allusions to finding members with complimentary thinking styles (see, e.g., Lewrick et al.'s (2018) rather pop-psychology prescriptions in their *Design Thinking Playbook*) which perhaps encourages a superficial acceptance and application of design techniques.

Design thinking in organizations

The emergence of design thinking as an all-encompassing method for generating new outcomes and value for organizations has been rapid. Touted as a general method for innovation, design thinking has now gained currency among organizational theorists and popular culture as a means of harnessing creativity to serve business ends. Writing in *Wired* magazine in 2014, Design executive Jeffrey Tjendra argued that design thinking was a necessary response to the logic-dominated educational system of 20th century America, which led to companies whose leadership were distrustful of creative thinkers. Thus, as technologies started to shift dramatically, industry needed a way to harness value from changing consumer demands and behaviors. By rendering the design-oriented problem-solving process more comprehensible to managers and accountants, design thinking offered a somewhat orderly process that could be managed, understood without specialized knowledge, producing outcomes that could be assessed in terms of value (and made a leap into popular acceptance that UCD has so far failed to do).

Over the last decade design thinking has moved from emphasizing product design to include all forms of services, and is touted frequently as a way for any organization to harness the power of 'out of the box' thinking to solve difficult, or in the rhetoric of design thinking, 'wicked' problems (Buchanan, 1992). It is frequently applied in education, healthcare, business, and manufacturing as part of efforts to improve processes, to engage participants and customers, and to encourage change management. As such, it is less of a formal method and more of an approach to targeted problem solving based on imagination, a philosophy of trial and error, and collective input.

Of course, as design thinking has been advocated for and adopted by diverse practitioner communities, it has been criticized as over-hyped and becoming part of the business school orthodoxy that pushes it as a solution rather than a form of reasoning and thinking (Wylie, 2017). Some critics go further, with US-based graphic designer Natasha Jen (2017) famously presenting a talk entitled *'Design Thinking is Bullshit'* at that year's 99u Conference, focusing her concern on the weak level of informed criticism in the outputs of typical design thinking exercises. While the general understanding of design thinking involves a stage of evaluation, Jen thinks this process is too weak, reliant on 'Post-it note' reactions and uninformed input from participants, stemming from the philosophy within design thinking that 'everyone's a designer', part of a feel-good marketing of the approach aimed at selling the method to business communities lacking any design expertise, which as Jen notes, serves to diminish the professional value of trained designers.

Related criticisms have been made of the claims for design thinking's impact, and the lack of a theoretical basis for informing decisions throughout the process. Case studies demonstrating value are rare, most justifications for the use of the method residing on quick reactions from participants or accepting clients who welcome the resulting 'solutions'. Norman (2018) argues that what really matters for rational justification is some form of comparison with other methods but this is difficult to do and while the criticism is valid, it is hardly unique to design thinking, it could be leveled against any design approach that does not use controlled comparisons to determine quality and long-term impact of designs. The absence of theory serves the inclusiveness of design thinking's approach but in so doing, it suggests that all inputs are equally valid and offers little assurance that insight rather than popularity guides the outcome.

Hyped or not, there is no denying that design thinking has raised awareness among many that creative solutions might best be enabled by embracing input and thinking broadly about the experience of users or others impacted by a designed product or service. Design thinking really is not a guarantee of successful outputs in all areas of application, and it is hardly a unique formulation of how groups might problem-solve or develop new ideas but in placing a strong emphasis on users and user-experience at emotional as well as behavioral or rational levels, it aligns well with the true spirit of user-centered design as traditionally understood and practiced by UX and human factors professionals for many years. It is this relationship that warrants further explanation as we believe that the justifiable criticisms of design thinking's lack of rigor or its favoring of impressions over theory, and thus its heavy reliance on the actual abilities of the individual team members, could be addressed through a stronger application of the theory and method bases of user-centered design.

What do we mean by UCD?

User-centered design (UCD) is employed here as an umbrella term for the various approaches that have emerged in related fields concerning the development and implementation of information technologies or systems that support human use. Ritter (2014) notes that various disciplines, from computer science to sociology, have shared in the development of ideas, methods, tools and studies that share a common aim but often described their approach with different terms. Hence, the general use of UCD covers labels such as User Experience (UX) design, human-centered systems design (HCSD), human-computer interaction (HCI), Interaction Design (IxD) and so forth. While one could draw historical distinctions in terms of origin or theoretical framing, these labels refer to a common view that in order to create usable and acceptable IT, we need to base our work on some understanding of how people think and behave in context.

The intellectual origins of UCD are key to understanding how we feel it can combine productively with design thinking. Ritter's overview lists 11 disciplines as foundational:

- Cognitive and social psychology
- Linguistics
- Mathematics
- Computer science
- Engineering
- Human factors and ergonomics
- Socio-technical systems design
- Scientific management
- Work, industrial, and occupational psychology
- Human relations
- Organizational behavior.

Further, she acknowledges that there have also been significant contributions from anthropology, information science and sociology as technologies have reached more of our world. Thus, at least she acknowledges 14 contributory disciplines, and there's an argument to be made that even that does not reflect every source since no mention is made of graphic design or arts, typography, and so forth, which all affect the design of user experiences. The precision of the listing however is less important than the diversity, which in turn suggests UCD can and does draw on deep intellectual traditions that have emphasized theoretical and empirical analysis of human activities. Consequently, there are libraries of research studies on user responses to designs, the impact of new technologies on organizations, the techniques that yield faster or cheaper testing results, or the manner in which user characteristics such as attitude, age or ability affect use of information systems.

It is this knowledge base that students of UCD study and gain familiarity with during their education. When coupled with appropriate methodological skills that allow them to gather and analyze data from users in a reliable and valid manner, UCD professionals seek to shape design and implementation processes to ensure the intended outcome for users. In the following section we extend this to explain further the practice of UCD.

UCD as an evolving set of challenges

It is common to talk about three waves or paradigms in the history of UCD or HCI since researchers Harrison et al (2007) suggested that concerns with the user-interface (first wave, which we might date approximately from the late 1950s to early 1970s) were broadened to include a deeper understanding of the cognitive dynamics of use (second wave, from the 1970s to mid-1980s) before a further broadening by the embrace of social and contextual analyses of situated action (third wave, 1990s-onward) came to dominate. One could argue with this telling or ask if we have we seen a fourth wave, particularly with the embrace of emotion as an important aspect of human-computer interaction in the past decade, but as a broad narrative it serves to highlight the progression of UCD over time. Rather than being a fixed methodology with a unified model of use and users through time, UCD has evolved as technologies changed, maintaining some key elements of focus (such as the empirical study of users) while recognizing some aspects of interactions warranting more or less attention than others (e.g., observing interaction at group rather than individual levels).

One could also take issue with the characterization of each wave as a particular paradigm or addressing a set of unique concerns that proponents at an earlier wave are unable to address. Even as we see a shift from single user interface issues in office work spaces to distributed or group activities in mobile environments, we cannot ignore the first wave concerns as we address the third wave design challenges but we must consider each level as concurrently part of the design challenge. For example, there is no less concern now with visual layout or physical ergonomics (classic first paradigm issues) in a world of where users have multiple small screens and deal with cloud-based services (third paradigm interaction patterns). In effect, UCD professionals rarely limit their work or scope of influence to a particular paradigm, as labelled by Harrison, though there are specialist exceptions. More likely, UCD professionals apply a knowledge base and a set of methods that is more universal and crosses supposed paradigms, as needed.

What is the theory and method base of UCD?

The three paradigm framing of HCI as indicative of the theory and method base underlying the work of UCD professionals. By this characterization we would invoke psychophysiology as the basis of much work during the first wave where concerns with physical ergonomics, workstation layout and task design dominated. Such early work drew heavily on the existing knowledge base of anthropometrics, reaction time, physical stress and workload, to help design a physical interface that improved human comfort and lessened repetitive strain. A good summary of work from that era can be found in Murrell (1965).

Similarly, we can characterize the second wave of HCI as dominated by a deeper focus on the underlying cognition of users necessitated by the increasing complexity of digital technologies and a commensurate shift in human working practices from physical to mental labor. As computers came to mediate language and visual information intensive tasks rather than solely numerical processes, more diverse users were engaged and design sought to understand how people perceived, processed and provided input to digital devices (see, e.g., Dillon, 2003 for a summary). Focus turned to navigation structures, menu design, input device response, and error correction as a new wave of technologies emerged. Card et al. (1984) delivered perhaps the most canonical work in this period with their formal model of human information processing derived from two decades of work in cognitive psychology to generate the calculation of performance times based on the analysis of goals, operators, methods and selection rules applied by humans in routine information tasks.

With the relentless evolution and adoption of technology pushing use into team and organizational work, UCD began to shift focus in the third wave toward concerns with collaboration, computer-mediated communication and shared activities, asking questions of human action that were not easily reduced to individual cognition on routinized tasks. Though not strictly speaking a shift to group analysis as much as a concern with deeper, situated analysis of the context and meaning of work, this approach acknowledges the complexity and importance of context in giving meaning to human activities in what has been described as a phenomenological turn. Harrison et al. (2007) considered this only an emerging third paradigm but it would seem to have closer ties to the philosophy of modern user experience and design thinking approaches than either previous wave, perhaps indicating that it is now fully extant.

As the attention of UCD researchers expanded, and with it a tendency to lean more or less heavily, if not exclusively, on dominant theoretical frameworks at each point, we can also trace an evolution in the methodological approaches of UCD professionals. While it's not as simple to draw firm distinctions across each wave or paradigm, the first two eras certainly placed more emphasis on classical scientific methods than might be the case in contemporary work. Partly due to a desire among the UCD community for recognition, the goals of providing reproducible results, clear evidence for design choices, and ultimately a verified theoretical set of standards that could be applied universally, made for a dominant experimental methodology in design. Prototypes were to be tested on real users. Final designs were evaluated in terms of effectiveness and efficiency of task performance, with UCD practitioners supposed to make recommendations for further improvement based on the detailed, often statistical analysis of real user performance data. As advances in cognitive psychology were being made in concert with new tools for measuring human information processing, scientific approaches in HCI reflected the dominant methodological paradigm of the era, and the legacy of experimental methods in data capture and analysis remains strong in contemporary evaluations.

However, there has always been more to UCD than laboratory-based evaluations and data capture. Use of surveys, interview techniques and task observation have a long history in the field and proved necessary for work at the earlier stages of design where user requirements were being captured. As the third wave of UCD emerged there was an extension of methodological coverage rather than a revolution. While studies of the more affective and emotional responses to design continued to analyze data at the single user level, the incorporation of group and organizational level analyses encouraged use of more field research methods drawn from anthropology and social psychology. A move toward conversation analysis, turn-taking in mediated communication, and situation awareness led to the adoption of ethnographic methods by many in the UDC community (a good example of such an approach can be found in Beyer and Holtzblatt (1998).

Surveying the practice of UCD now, we see a plurality of methods from the history of the HCI field used, both formal and informal, as user-centered efforts seek to influence the design process at all stages. While practitioners vary in their preferences and abilities, it is commonly expected that a UCD professional is competent in a variety of data solicitation methods, from observation to interview techniques to test design and statistical analysis. It is this reliance on evaluation methods in particular that signal a distinction with design thinking, but we should also acknowledge the theoretical leanings of UCD practitioners, usually drawing principally from cognitive or from social epistemologies, as a further marker of distinctiveness between these two communities.

UCD in action

While the model of design thinking presented above might superficially appear familiar to many practitioners of user-centered design, with its emphasis on requirements, prototyping and testing, there are significant differences between these two approaches. The main differences spring from the educational background of the participants in each. Traditionally practitioners in UCSD earned human factors or psychology degrees, often treated as a requirement for employment. A substantial part of this type of education is the focus on research methods, psychometrics and user psychology. A voluminous body of knowledge about human cognitive, behavioral, social, emotional capabilities, limitations and proclivities is leveraged within the human factors literature to inform the design of systems, processes and products.

For human factors researchers in the UCD tradition, the ISO-9241 and related standards for usability engineering, with their emphasis on operational characteristics such as effectiveness, efficiency and satisfaction, are consistently used to frame both formative and summative testing. Combined with knowledge of psychometrics and measurement theory, usability criteria related to user engagement with systems can be accurately benchmarked. UCD professionals also employ structured evaluation methods for prototypes such as the cognitive walkthrough technique which offers a lens on the breakdowns in communication between a user and interface to reveal problems that quick inspections can easily miss.

In contrast to the 'bias for action' mantra of initial DT practice, UCD tends to place heavy emphasis at the earliest stage of a project on understanding as much as possible about users and the contexts in which they will be engaging with the design. Employing an array of qualitative research methods such as case study, ethnography, or contextual enquiry in the discovery phase of research, UCD is particularly systematic in identifying user requirements and even ensuring their appropriate participation on the design team.

The skills types, methods and educational background of UCD professionals reflects a different orientation than DT, and seems to offer what might seem to be barrier to involvement that is not present in the type of workshop activities encouraged by the design thinking process. Both approaches can be seen to offer particular advantages (UCD has depth, data, and methodological rigor to build confidence in decisions; DT enables every stakeholder, if involved, to engage in solution generations). The ideal approach might help to blend the strengths of each.

Framing the comparison in the context of design acts

The greater structure and emphasis on theory and methods in UCD can lead some people to view it as cumbersome or outdated compared to the more agile approach of design thinking. Some people have presented the distinction in terms of scale and aim or in terms of origins and intellectual roots, e.g., Nallan and Jaiswal (2019) acknowledge that both UCD and DT approaches share common goals but they view design thinking as intending to influence broader strategic thinking, innovation and business repurposing. Certainly, UCD's origins in social science and engineering give it the theoretical and methodological character it has, and design thinking often presents its own origin story as a reaction to traditional approaches. However, rather than frame the discussion here through a compare and contrast exercise, or as a right or wrong analysis, we believe the best way to understand the two approaches in action is to stand back and examine the nature of design more generally, in particular to understand how the inputs of each approach can enable and encourage design solutions and interventions that matter for people, particularly in the context of information systems design.

The study of design

Both design thinking and user-centered design are derived in an intellectual sense from the broader area of design studies. Although design has been a human activity for centuries, it has only recently been codified and studied in a form that suggest it might form a unique or recognized discipline. There are few schools or departments of design in academia outside of specialized areas within engineering, architecture, or art disciplines, testimony to the view of design as a process that exists across application areas rather than as a unique intellectual focus. Nevertheless, the view of design as a form of human activity that warrants serious investigation is older than many appreciate.

American scholars Rogers and Bremner (2017) argue that the design used to be appended to specific areas of application, e.g., product design, visual design, textile design etc. However, since the 1950s these discrete application areas have dissolved to create "more fluid, evolving patterns of practice that regularly traverse, transcend and transfigure disciplinary and conceptual boundaries" (p.22).

Lawson (2006) traces efforts to conceptualize design in almost scientific terms back to the 1920s, when architects sought to rationalize and objectively demonstrate a system for designing. A further flourishing of interest in studying design methods in the 1960s crystallized this push to 'scientise' design (Lawson, 2006, p.94). The *Conference on Design Methods* of 1962 is considered a turning point in this move to develop design as a distinct discipline. Though mainly originating in architectural and industrial design literatures, the interest in design as a human form of problem solving worthy of serious study was also advanced within cognitive science where Herb Simon, in his seminal work *The Sciences of the Artificial* (Simon, 1969) advocated for design to be formalized as a discipline within the academy. Speaking directly to the problem Simon issued a call to develop better education around the science of design to encapsulate .... "a body of intellectually tough, analytic, partly formalizable, partly empirical, teachable doctrine about the design process" (p.113).

Design Studies, a leading scholarly journal with the stated aim of 'developing understanding of design processes' across disciplines, was first published in 1979 and continues to this day. The first issue made a case for treating design as discipline, forming a better understanding of what Lawson (2006) calls 'design in general' from the intersection of studies of design processes to ways of thinking and problem solving. Now, there are more than 20 scholarly journals regularly publishing refereed papers on aspects of design drawing authors from across the design world and academia, suggesting that as a discipline, design has developed an identity.

Despite the scientizing attempts, many still distinguish between scientific thinking as a pattern of problem-solving behavior aimed at determining the relationships between things that exist, and design thinking, as Simon puts it, which focuses on what ought to be. This certainly seems more meaningful a distinction than one that pushes science as purely rational and design as more intuitive, a somewhat cliched framing that is often found in popular descriptions. While it is clear there are both similarities and distinctions between design and science, the issue might be exaggerated to defend intellectual traditions or practices. For example, Charles Owen (2007), a professor of design at the Illinois Institute of Technology, states that design thinking is the obverse of scientific thinking. He describes two ways of problem solving: Finding and Making (design); and Analysis and Synthesis (science). He posits that design as a field is at the opposite end of the spectrum from fields characterized by analytic activities and processes.

While this offers a neat simplification, it does not cleanly fit with the majority of the research studies on designers and their problem-solving processes which tend to show strong similarity between scientific and design problems solving. Indeed, one of the most famous formulations of architectural problem solving by Darke (1979) utilizes classic scientific terms of conjecture and analysis to describe many top designers' processes. Lawson's classic study comparing designers and scientists trying to solve a problem emphasized a distinction built around emphasis on solutions (designers) and processes (scientists) but a common commitment to understanding the problem.

Tim Brown (2009), leading thinker in the IDEO design group has described design thinking as perhaps a 'third way' that sits between intuition and rationality, which has superficial appeal that perhaps acknowledges both sides but leaves open the question of how to overcome the continuing challenge of our reliance on individual skill or intuition in managing design activities at scale. There may indeed be great designers, and some may operate in that space between intuition and rationality, but there are rarely enough of them to tackle the number of design challenges we face in our increasingly designed world. To this end, we may wish for greater rationality so as to educate more skilled design practitioners, or at least to provide the theoretical and methodological scaffolding to address the challenges.

There seems to be an inherent resistance in some areas of design to answering Simon's call for greater formalization of design knowledge and education. Cross (2001) argues that for some, 'designerly ways of knowing' do not want to be influenced by scientific thinking. Alexander (1971) captures this succinctly in saying "the ultimate object of design is form", which is much like scientists claiming their ultimate object is theory, but education in the sciences is not primarily about making each student a theorist as much as a methodologist, and there might be a useful lesson here for the design thinking movement as we seek to advance the practice of design to serve all humans better.

Reconciling UCD and design thinking approaches

The classic design thinking process shown earlier lacks rigor in particularly key areas. The empathy stage tends to emphasize quick analyses of the context in which users are situated and little formal checking of data and assumptions. Users are presented as personas and in their education, which invariably lacks formal classes in user analysis techniques, and design thinkers are often encouraged to use story-telling as a way of summarizing user activities and needs. While such narratives might be useful summarizing and presentation forms, they can easily lack any comparisons with the reality users experience, and design thinking as a method seems to eschew one of the most basic rules that scientists learn i.e., the practice of comparing what you think you know with what actually happens so as to identify and overcome bias.

Decades of psychological research confirms that humans are quick to fill in the blanks, to assume patterns and to confirm their impressions with very selective data sampling from the world. This pattern is a function of our own cognitive system dealing with continual sensory data overload and a desire to free up our mental resources. While our minds are remarkable, we are easily seduced by patterns (real or imagined) and tend to automatically fill in gaps in our understanding by drawing partial conclusions and using these as basis to proceed to handle new situations. Daniel Kahneman (2012), an Israeli psychologist, calls this 'System 1' thinking, the automatic cognitive processing we all engage in when handling the continual sensory input of the world, and he presents myriad examples of the fallacies humans routinely adhere to in the face of alternative data, information or evidence. Our cognitive habits as humans lead us to infer rapidly, quickly leap to conclusions, react automatically and 'trust our guts'. Much of the time this serves us well in the world where we need to think fast in our routine interactions but it encourages us to build coherence among ideas that might not actually have evidence, to rely on our biases and to engage in a in a type of story telling about the world. In contrast, Kahneman's 'System 2' style of thinking is deliberative, analytical and controlled, utilizing reasoning, memory and existing knowledge to help us draw inferences and understand the operations of the world around us. This form of controlled processing underlies knowledge-based activities but it comes with a cognitive cost in terms of speed and efficiency.

Kahneman shows through countless studies that our natural response to the world is not to process data and determine the facts, as we might imagine we are doing, but rather to reach a conclusion first and then to adopt supporting arguments. We as humans have evolved to think 'fast' for survival. However, truth and validity are not well served by 'System 1' thinking. Our 'System 2' checks for validity and rationalizes conclusions, but this is not an automatic process, it demands effort and time. The scientific process is in many ways a recognition of this and a representation of problem solving that forces a 'System 2' style of thinking. Design thinking makes no allusion to this and as such, its action-first disposition can easily lead to poor or limited impressions of user needs that are not checked or balanced systematically against the real world. UCD has, for many years now, understood the complexities of real-world usage and the problems that are associated with poor analyses of contexts, tasks, and user groups. Indeed, for many user-centered designers, the efforts put into real analyses at this stage of a design often exceeds the time and effort put into testing and other phases.

A second area of possible improvement for design thinking that UCD offers is in the evaluation methods that can be employed at the testing phase. In design thinking accounts, impressionistic and informal testing seems to dominate practice. UCD emphasizes the importance of impartial feedback from representative users. Ideally, designers do not test their own designs since they are obviously invested in them. Jakob Nielsen (2007), the UX author and consultant, affirms that utilizing specialized UCD evaluators, is important for many reasons, e.g., evaluators have specialized research skills and understand test procedures, they are more objective to hearing negative feedback, are less biased in selecting an appropriate range of tasks, and usually are trained in data treatment and analysis.

With early low fidelity prototypes researchers will conduct formative testing that aims to elicit design direction. With later higher fidelity prototype testing the purpose is to begin to validate the design concept against intended outcomes. The methodologies associated with different testing objectives varies and an experienced researcher will develop appropriate test protocols for each. UCD research has also uncovered biases in test processes, determined the impact and confidence levels for sample sizes, studied how items and test prompts can influence user responses and so forth. Indeed, some of the major advances in user-centered design practices have come from the scientific study of test methodologies and tools, and this knowledge is part of the curriculum in most if not all curricula in UCD education. Again, this is a form of knowledge that would likely strengthen design thinking if more of its practitioners were educated in methods.

There are however lessons for UCD that design thinking can provide. One of the great successes of the design thinking movement has been its democratizing of input to design. The type of design thinking practiced in industry has few barriers on participation, anyone in a group can engage without needing technical skills or design experience. Ideas, at least at the divergent ideation phase, are encouraged not criticized and through this, engagement with a design process or organizational change issue is enhanced (though we should note, this is more often claimed than objectively assessed).

Specifically, the idea of design as an act to which anyone can contribute removes some of the mystery of the creative process underlying design. Even though early studies on designers sought to counter the belief that only certain people possessed the unique abilities to design, the myth remained and the phrase 'starchitect' is often used to refer to celebrity status of acclaimed architects and certain designers become famously associated with distinct products such as the iPhone or consumer electronics. The view that these individuals were uniquely skilled or created a product much the way an artist delivers a masterpiece has proven difficult to shake from popular culture. Within a design thinking workshop, the process is intentionally more collective, less the result of genius than the consensus output of idea pooling and free expression.

In this way, one might argue that design thinking is itself more user-centered, in as much as the participants in the in the process face few barriers to entry and the process tends to pursue an egalitarian level of input. Consequently, getting participant buy-in to DT methods seems easier than with some methods of UCD, such as early and repeated user testing with adequately representative users. Given the well-documented problems of UCD methods gaining sufficient traction in large scale design processes, UCD practitioners might benefit from some of the lessons of design thinking regarding involvement and selling of the method to outsiders.

Our belief is that by adopting the strengths of both approaches, the rigor and knowledge enshrined in UCD and the flexibility and easy participation of stakeholders in design workshops advanced by DT, there is a way of advancing true user-centered design practices more widely. Studies of designers, both in situ and in lab-based experiments, have been conducted since the 1970s, and this work has informed the HCI community's own efforts at understanding how information artefacts are created and how we can improve the process. It is plausible to argue that the whole user-centered movement was both motivated by and in turn itself motivated further studies of designers in an effort to codify methods that could improve the quality of user interfaces, and smooth the organizational adoption of new information technologies through better design and design processes.

Conclusion

We might usefully acknowledge that UCD professionals and design thinkers are involved in a common enterprise, one that is part of the broader, ongoing and continually evolving emergence of design as a discipline. The pace of disciplinary emergence is relatively slow, spanning decades or more, and often involving generations of scholars and practitioners. In this light, we might do best to recognize design thinking less as an alternative to UCD, or as a competitor for attention than as a manifestation of the general desire to find ways of shaping the design process to serve a shared goal: the provision of more acceptable and usable technological artefacts.

We believe design thinking could be improved, methodologically, by incorporating greater understanding of the research base of UCD, in particular the deep knowledge that has been obtained of user psychology, as well as the value of evaluation methods in generating valid and reliable estimates of usability and user experience with a design. Similarly, we believe UCD could gain value from the methods Design thinking uses to engage people in the design process actively and constructively.

What would be required for this to happen is a shift in the educational process for both sets of practitioners. We have not explored this issue here but it is likely the lack of awareness of user psychology and evaluation methods in much design education is rooted in disciplinary differences which might not be easily addressed through short courses or the addition of one class to a degree program. In many ways, we are still witnessing the birth of a design discipline as imagined by Herb Simon.

References

Alexander, C. 1971. Notes on the Synthesis of Form. Boston: Harvard University Press.
Beyer, H. and Holtzblatt, K. 1998. Contextual Design: Defining Customer-Centered Systems. San Francisco: Morgan Kaufmann.
Brown, T. 2009. Change by Design How Design thinking Transforms Organizations and Inspires Innovation. New York: Harper Business.
Buchanan, R. 1992. Wicked problems in design thinking. Design Issues, 20(2): 5–21.
Card, S., Newell, A. and Moran, T. 1984. The Psychology of Human Computer Interaction. London: Ablex.
Cross, N. 2001. Designerly ways of knowing: Design discipline versus design science. Design Issues, 17(3): 49–55.
Darke, J. 1979. The primary generator and the design process. Design Studies, 1(1): 36–34.
Dillon, A. 2003. User interface design. pp. 453–458. *In*: Nadel, L. [ed.]. MacMillan Encyclopedia of Cognitive Science, Vol. 4, London: MacMillan.
Dourish, P. 2004. What we talk about when we talk about context. Personal Ubiquitous Computing, 8(1) (Feb. 2004): 19–30.
Gardiner, M. and Christie, B. 1987. Applying Cognitive Psychology to User Interface Design. Chichester UK: Wiley.
Harrison, S., Tatar, D. and Sengers, P. 2007. The Three Paradigms of HCI. Proceedings of ACM SIGCHI 2007, New York: ACM Press.
Nallan, H. and Jaiswal, M. 2019. UCD v. Design Thinking, https://think.design/blog/ucd-vs-design-thinking/ June 2019. Downloaded June 11th 2020.

Jen, N. 2017. Design Thinking is Bullshit. https://99u.adobe.com/videos/55967/natasha-jen-design-thinking-is-bullshit.
Kahneman, D. 2012. Of 2 Minds: How Fast and Slow Thinking Shape Perception and Choice. https://www.scientificamerican.com/article/kahneman-excerpt-thinking-fast-and-slow/.
Lawson, V. 2006. How Designers Think: The Design Process Demystified. 4th ed. New York: Architectural Press.
Lewrick, M., Link, P. and Leifer, L. 2018. The Design Thinking Playbook: Mindful Digital Transformation of Teams, Products, Services, Businesses and Ecosystems. Hoboken N.J.: Wiley.
Murrell, K.F.H. 1965. Ergonomics: Man in his Working Environment. London: Chapman and Hall.
Nielsen, J. 2007. Should Designers and Developers Do Usability? https://www.nngroup.com/articles/designers-developers-doing-usability/ Article downloaded 8/10/2020.
Norman, D. 2018. Rethinking Design Thinking. https://jnd.org/rehtinking_design_thnking/Article downloaded 12/10/20.
Owen, C. 2007. Design thinking: Notes on its nature and use. Design Research Quarterly, 2(1): 16–27
Ritter, F., Baxter, G. and Churchill, E. 2014. User centered system design: A brief history. pp. 33–54. *In*: Ritter, F. Baxter, G. and Churchill, E. [eds.]. Foundations for Designing User-Centered Systems What System Designers Need to Know about People London: Springer-Verlag.
Rogers, P. and Bremner, C. 2017. The concept of the design discipline, Dialectic, 1,1, Winter DOI: http://dx.doi.org/10.3998/dialectic.14932326.0001.104.
Shackel, B. 1959. Ergonomics for a computer. Design, 120: 36–39.
Shapiro, E. 2020. Is Design thinking really BS? A virtual debate between Natasha Jen and Marty Neumeier. UX Collective Jan 2020. https://uxdesign.cc/is-design-thinking-really-bs-5deb6c333f2.
Simon, H. 1969. The Sciences of the Artificial. Boston: MIT Press.
Sproull, B. 2018. The Problem-Solving, Problem-Prevention, and Decision-Making Guide. Organized and Systematic Roadmaps for Managers. New York: Routledge.
Von Thienen, J., Clancy, W.J. and Meinel, Ch. 2019. Theoretical foundations of Design Thinking. Design Thinking Research, Design Thinking Research Looking Further: Design Thinking Beyond Solution-Fixation, Ed. Christoph Meine Larry Leifer, Springer Cham.
Tjendra, J. 2014. The origins of Design thinking. Wired. https://www.wired.com/insights/2014/04/origins-design-thinking/Downloaded Dec 10th 2020.
Wylie, I. 2017. Design Thinking: Does it live up to the hype. Financial Times, Oct 11th 2017 https://www.ft.com/content/a961cada-a520-11e7-8d56-98a09be71849 Downloaded Dec 11th 2020.

CHAPTER-4
Situated Design Thinking
Åsa Wikberg Nilsson

Introduction

Design is deeply integrated into everyday human lives. On an ordinary day, some people may be interacting with design(s) through the visual communication of advertisements in the subway, by adjusting the car seat, through interaction with a mobile app, or by finding their way in the city. In everything from simple fragments to complex constructions in human environments, there are visual, material, and immaterial design representations in signs, objects, interactions, systems and environments. We humans are design users, eagerly or reluctantly, and use design(s) to shape, represent and make sense of our individual and collective existences (Dilnot, 1982). The concepts of *design* are referred to here as the practice and process of designing, *designs* as the result of such processes, and *design(s)* for indicating both practice/process and its results.

Design thinking has in the last decade or so spread almost as fast as the corona virus. American design consultant Tim Brown defines it as a creative approach to innovation that builds on design methods, and can be implemented for anything from products and services to solve complex social problems (Brown, 2008). In this sense it is rationalised as a more strategic role for designers, to create ideas and concepts that better solve users' and customers' needs. American design scholar Richard Buchanan describes design thinking as the neoteric art of design, i.e., a practice of new learning (Buchanan, 1992). In this sense, designers' skills lie in the ability to address wicked problems, which are open-ended and do not have a right or wrong solution. Linear models are clearly unfit to solve such problems. In Buchanan's (1999) view, there is a need for design thinking as the new learning of how to be "making products—and

---

Industrial design, Department of Social Sciences, Technology and Arts, Luleå University of Technology, Email: asa.wikberg-nilsson@ltu.se

by "product" I mean a range of phenomena that is very broad, including information, artefacts, activities, services, and policies, as well as systems and environments—[as] the connective activity that integrates knowledge from many fields for impact on how we live our lives".

In this text, the social and situated role of design is outlined. The material and immaterial designed landscapes we humans surround ourselves with affect our daily experiences: they affect our performance, our sense of well-being and our sense of social belonging. Good design can support and empower our lives. The opposite, however, may also ensue, i.e., designs that make us frustrated or even discriminates. The everyday people who face designs may define themselves—or be defined by others—as female, non-binary, senior, function-variated, coloured, or some other kind of human being, who have in different ways advantages, or disadvantages, from interrelating with design(s). Design practice is for this reason described by Swedish design scholars Yvonne Eriksson and Anette Göthlund (2004) as a discourse, which is based on certain conventions and stereotypes that express certain ideals, norms and values. The consequences of design thinking may hence from a critical standpoint be questioned, since it rarely outlines which social group the solutions are relevant for, or include analyses of situated power differences among designers and supposedly non-relevant social groups (Berg and Lie, 1995). This begs the question if designers are better at discussing the methods applied in design thinking, rather than what kind of thinking it is based on, and what consequences this results in for different people. How does design thinking motivate the human-centricity, i.e., what Buchanan (2001) states as the first design principle that design is fundamentally about human dignity and human rights?

The basis of this text is that design thinking needs to be critically explored in terms of social norms and biases embedded in representations, materials, actions, and thoughts, and as a subject of genuine user value, -dignity and -rights. As once stated by Austrian-American designer Viktor Papanek and American designer/architect Buckminster Fuller (1972) in terms of social and ecological design, there are professions that are more harmful than design, but only a few of them. In relation to the social aspect of design, tokenism, user insight drifts, and oppression (indeed), as in unawareness or reluctance of some use and user experiences, can be aspects that count into such malignance. In this text a need for new perspectives is proposed for counteracting the many inequalities that to some extent have their origins in design. Drawing on Buchanan's (1992) outline of four orders of design; symbols, objects, interactions and systems, are merged with the American philosopher Donna Haraway's (1988) concept of 'situatedness', i.e., awareness of meaning and representation in relation to participating actors, things, and context; the following sections outline what, why, and how, a framework of situated design thinking might contribute to norm-creative, and hence more inclusive, design(s).

Thinking of norms in design

American sociologist Jack P. Gibbs (1989) sees social norms as both visible and invisible attitudes, values and ideas that govern peoples' attitude and behaviour in relation to different things. Some examples of norms are social categorizations based on, e.g., gender, ethnicity, age, functional variation, and so forth, and how these groups are prioritized in for example design (Appleton-Dyer and Field, 2014; Wikberg Nilsson and Jahnke, 2018). One such norm is described by American sociologists Candace West and Don Zimmerman (1987) as the 'doing of gender', which incorporates socially and culturally constructed values and meanings associated with men/male and women/female in terms of power and subordination. In French writer, intellectual and feminist theorist Simone de Beauvoir's (1997) view, this involves hierarchies and valuation of bodies, behaviours, occupations, clothing, colours, forms, and so forth, as the norm, i.e., the natural, or the second, the deviance. Such norms can be seen as socially formed conventions that affect how we think of both people and things.

Here is an example of how common held norms of different social groups display themselves as stereotypes. Stereotypes of women involve them being passive and desirable, and of men being active and adventurous, but also aggressive and violent (Ruble, 1983). Although most people would agree that all men aren't violent, social norms contribute to some stereotypical representations seeming natural and universal. This can be explained, by American philosopher and gender theorist Judith Butler's (2011) notion of gender being performative: we tend to see what we expect, and we perform gender roles in relation to what is expected of us as women and men. Gender performance involves different social groups acting as directors and actors, as well as voluntary or involuntary spectators in the fiction that is played out in our everyday life. French philosopher Michel Foucault (1990) explicates the differences between biological and social gender, in the sense that social gender is a mean for creating both power and privilege, as well as discrimination and oppression. This involves new theories such as queer theory, i.e., the rejection of binary gender norms in society (Butler, 2011). Social norms are hence not shaped by individuals' wills or behaviours, but performed and lived in different cultures and contexts.

Social norms also affect power, representation and privileges for different people from an intersectional perspective (e.g., Crenshaw, 1989; Morgan, 1996; Berg, 2010). Racial experiences influence how we see the world, as American art and art pedagogy scholar NaJuana Lee (2012) points out: similar to gender, the concept of race is based on the idea that one group of people is distinct in terms of physical appearance, and a social convention of how that group thinks, acts, and what it values. American philosopher and critical theory scholar Kimberlé Chrenshaw for this reason developed the concept of intersectionality, with axes that define valued or devalued aspects in relation to other aspects, i.e., a black homosexual man and a white disabled woman and a non-binary senior person have different privileges and opportunities compared to a white middle-aged functionally-abled heterosexual man (the norm) (Crenshaw, 1989). Figure 1 exemplifies such intersectional axes based on different norms, privileges and oppression.

The Swedish gender and design scholar Magdalena Peterson McIntyre describes the doing of norms in design as a continuous silent dialogue between people and the material world, where design manifests attitudes and values: symbols, forms, colours, features, materials, and other design elements both explicitly and implicitly carry meaning and thus become part of a continuous doing of gender and other power relations in society (Peterson McIntyre, 2013). Design, like gender, is performative and relates

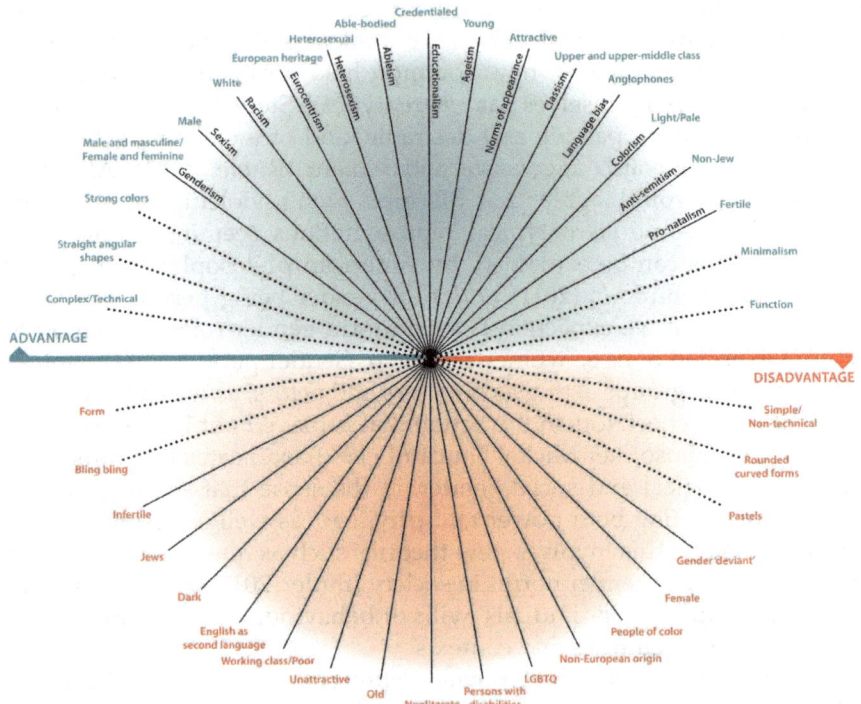

Figure 1. Intersectional axes of privilege and oppression based on norms, adapted from Morgan (1996), and with inspiration from (Ehrnberger et al., 2012; Alves Silva et al., 2016). Illustration: Åsa Wikberg Nilsson.

to wider social and cultural processes and conventions that construct certain meaning and value in the object itself (Peterson McIntyre, 2018). Overall people seem to associate gender with colours from a very early age, and boys and girls share similar gender stereotypes which seem to be inspired by media (Navarro et al., 2014).

British human geography scholar Suzanne Reimer (2016) emphasizes that design as an industry and profession should acknowledge that socio-cultural constructions of skill, craft and creativity is fundamental in understanding design thinking, actions and outcomes. In this sense, it is vital to address the power relations in design, where gendered norms appear almost taken for granted, marginalisation is created through dynamics of sexism, racism and discrimination by age and sexuality. For example, design professions such as advertising, product design and architecture remain white middle-class male-dominated sectors, which is reinforced by applying a seemingly gender-neutral understanding of learning, knowledge creation and social interactions. The reluctance of situated power relations and the creation of seemingly gender-neutral thoughts and practices are in this sense problematic, since they contribute in a re-construction of barriers to and in design(s).

British design historian Pat Kirkham (1996) outlines the deeply rooted power relations that design and gender are linked to, and states that these contribute to the fear of being perceived as female among men, as it means a loss of power that many cannot stand. A personal experience in relation to this is a father raging against the personnel at the maternity hospital because there were only pink baby cloths, no blue ones, and his new-born son could certainly not wear pink. What if the baby boy grows up loving pink, just imagine his father's rage, not to mention the child's distress of not satisfying his father's ideals. Most people would certainly rationally agree that colours cannot hold such attribute as being dangerous, but emotions and logic seem to be two different things in terms of social norms. Yvonne Eriksson and Anette Göthlund (2004) explain this being rooted in humans being surrounded by visual communication that represent more than the direct visceral experience, as meanings and representations are inscribed in a cultural and historical context. Visual communication in media for example tend to represent some social groups in simplified ways, which can influence our perception and behaviour towards that specific social group.

It is generally more acceptable for women to do something that is linked to a male gender domain, than the opposite. However, this sometimes requires a transformation of the designed object itself. For example, a power drill can be transformed into a "little pink tool", a simplified, minimized, and pink-coloured version, to clearly distinguish it from the original. British design historian Penny Sparke (2010) refers to this design strategy as 'shrink it and pink it', which exemplifies how design contributes in re-constructing norms, in which the woman/female is the secondary and simplified thing that needs to be clearly separated and distinguished from the norm. Similarly, Swedish design scholars Åsa Wikberg Nilsson and Marcus Jahnke (2018) describe a razor as a product that has a basic function to remove unwanted body hair. But the form can be transformed from straight, angular shapes, accompanied by strong colours, features, action elements, and names such as Mach^3 Innovation, to soft, simple, and pink-coloured rounded shapes with names such as Venus and Passion. The text connotation in combination with the symbols and forms is what French semiotician Roland Barthes (1993) refers to as 'anchorage': it directs the viewer to the signifiers of the image, by means of a beforehand chosen subtle meaning. Hence, the communication and the objects in themselves demonstrate and inscribe the stereotypical ideals of—and constructed antithesis between—women and men. The differences are reinforced, or anchored in Barthes words, by adding the text "for women" to the pink one, something the original does not seem to need.

Another example of norms in design is the microwave oven. British science and technology scholars Cynthia Cockburn and Susan Ormrod define that in terms of function, it is a product that heats food through rotating water molecules in food through microwaves. In terms of form, it however transformed from a brown-coloured high-tech product, to both a simpler form and function supplemented with a pastel tone, when the target group changed from men to women (Cockburn and Ormrod, 1993). These examples of design strategies seem to be of reinforcing a gender dichotomy, as there aren't any obvious differences in what the things are expected to do for their users, whether they are women, men, non-binary, black, pink, or any other colour.

Swedish design scholar Pelle Ehn clarifies that as humans we are surrounded by things that from one perspective can be seen just as a *denoted* physical device that gives access to some function or performance: a chair, a phone, or a building (Ehn, 2008). However, the same device also has a *connoted* symbolic representation that varies the user interaction and experience. Swedish cultural geography scholar Tora Friberg and Swedish architecture scholar Anita Larsson discuss public buildings and spaces, which in their view materialize patriarchal hierarchies and structures through re-constructing industrial revolution ideals such as concrete, stone, and glass (Friberg and Larsson, 1999). A city hall, for example, can have dark oak walls, heavy doors and golden frames filled with photos or paintings of the men of power. This represents a power representation in which women, non-binary, and non-whites are excluded, in which men have created their own gender domain. Similarly, our homes are not neutral, they are not only a matter of building technology (material phenomenon), but also convey social and cultural norms of family patterns (immaterial phenomenon). This means living in a socio-material world, a world of design(s), which mediates our bodies' position in the world and our ability to take part on equal terms.

Norms in design might be explained through French science and technology scholar Madeleine Akrich's metaphor of the design process being similar to creating 'scripts', just as in a movie script and just as a film director: the designer creates action scenarios that guide the user to a certain behaviour, role and interaction through the denoted and connoted representation (Akrich, 1992). Akrich highlights the risk of unaware and systematically executed design practices being based on an 'I methodology', i.e., an un-reflected non-situated practice, which may result in certain persons in-scribing, i.e., having the privileges to interact with the design object, while others are de-scribing, i.e., cannot or will not interact (Akrich, 1995). American sociologist Susan Leigh Star (1991) illustrates this phenomenon through describing a designed thing that designers think does its job well, provided that you put aside maintenance and all of the user sectors that are discriminated against. Star for this reason raises the question of *cui bono*? Who benefits? One might also add; how does this align to the stated human-centricity in design thinking?

To explore possible answers for these questions, designers can also be seen as actors in the process. They are performing design practice as a form of scripts, through both physical and mental routines to which they inscribe, whether learned in design education, cultivated in design practice, or reinforced by societal norms as a whole. In line with German sociologist and cultural theorist Andreas Reckwitz, designers do their things as individuals, but the actions are also developed as a social routine, which become part of supplementing the design discourse as a whole (Reckwitz, 2002). This can be exemplified by Dutch-Australian design scholar Kees Dorst statement that designers create the environment in which they operate, they create their approaches to design situations, the roles they have in design projects, the coalitions they work in, how they approach users, and so forth (Dorst, 2008). There are also studies that consider the influence of the designer's individual attitude, values and design philosophy and how the social world of design organization, management and company policies contributes in shaping the design(s) (Pile, 1979). American sociologist Harvey Molotoch describes this as designers' narratives, assumptions, and expectations as being embodied in the things they produce (Molotoch, 2003). In this sense, companies, designers, traders, advertisers, and other middle-men impact what a thing can be and how it is made. Better understanding of the "stuff" system, including the mechanisms of change and stability, might improve designs, both socially and ecologically, in combination with exposing the stories in society which has created them.

British-Finish design scholar Guy Julier portrays the design thinking approach as contributing to a restating of design practice, in which designers are not only seen as creators of form, but also act as cultural intermediaries (Julier, 2005, 2008). Hence, design involves more than creating a form, meeting a representation, or transmitting a message to be seen, read, used, or otherwise interacted with: it includes structuring entire systems we humans face and interact with through designers' visualised and materialised world.

Thinking of knowing in design

Design theorists Horst Rittel and Melvin Webber describe design situations as 'wicked problems', i.e., problems which cannot be definitively described, that are not objective, which cannot be meaningfully answered as true or false, and which hold no neutral solutions that work in all situations (Rittel and Webber, 1973), that is, they are always situated. This can be challenging, and knowing becomes a completely different subject compared to finding the answers in a book beforehand and then knowing the right solution to the situation at hand. In such cases, following a strict solution-focused design model might be tempting, as it may seem to take away accountability for all connections and consequences in the wicked situation. However, instead of not tackling the wickedness, there are ways of organizing the knowing.

One way is to learn more of how people think. There are different ways of thinking. For example, American design theorist Don Norman discussion of individual cognitive schemes is interesting for thinking of how we might challenge ways of doing things (Norman, 2003; Norman et al., 2003). These cognitive schemes consist of *visceral*, as the automatic, intuitive, rapid judgment of what is good or bad, safe or dangerous, *behavioural*, as the routine processes that control everyday behaviour, and the introspective level of *reflection*. On a basic level, it relates to thinking about what we see, do and reflect upon. This also relates to Israeli psychologist and Nobel Laurate in economics Daniel Kahneman's discussion of bias when people assume something, as part of what is referred to as system 1 and 2-thinking (Kahneman, 2013; Morewedge and Kahneman, 2010). The kind of thinking that System 1 uses is in this sense automatically and intuitively generated representations, based on prior knowledge, background, values and so forth. The more reflective and energy-consuming system 2-thinking only starts when system 1 cannot process the challenge. This also includes anchoring, as when peoples' starting point, e.g. the formulation of a design problem, uses system 1-thinking and hence not challenge own beliefs, prior understandings, and the actors' own judgmental heuristics and biases (Tversky and Kahneman, 1974), hence human behaviour is influenced by the current context and norms within. Kahneman argues that many people overestimate that they know things, when they actually don't. Also, in ambiguous situations, people tend to see, think and act in a manner that is consistent with their expectations, or schemas (Fiske and Taylor, 1991). It is sometimes said that we see what we expect to see. Thus, people tend to see, act and think of things without the energy-absorbing system 2 kind of thinking, and will continue to do so, as long as nothing disrupts such ways of [design] thinking.

According to American philosopher and design theorist Donald Schön, thinking in design requires an active conversation with the material a situation constitutes, to be reframed into knowing-in-action (Schön, 1985, 1995). This involves a questioning of structural assumptions, a restructuring of strategies, and developing an understanding of particular situated phenomena. This action theory comprises two separate elements: *espoused theory*, i.e., the actor's idea of how they think they would act, and *theory-in-use* that actually governs how they behave (Argyris and Schön, 1975). The core of this theory is that it is neither enough to ask a user, nor a designer, how they would behave in a particular situation. To understand their behaviour, one should instead observe a larger and wider context than the concrete situation that one intends to transform.

American economist Richard Thaler and law scholar Cass Sunstein outlines a theory of nudging, that is, how to use design to guide people into what is seen as better choices, for example healthier, more sustainable and so forth. This theory is interesting as it involves an understanding of how people generally act, i.e., mechanically rather than knowingly, intuitively rather than sensibly, and through designs guiding people toward what is seen as better actions (Thaler and Sunstein, 2009). In this sense, each and every day we humans are subject to various biases that makes us take non-rational decisions. By knowing more about how people think, thoughtful design can nudge people to more valuable options, without restricting their freedom of choice. Nudging can hence be seen as a tool to support decision-making in complex situations, without having to apply system-2 thinking. For example, wayfinding symbols such as arrows on the ground guide people in certain directions. American design scholars William Lidwell, Kristina Holden and Jill Butler describe nudging as a design method for choice architecture, referring to the fact that people prefer the easiest way when taking decisions (Lidwell et al., 2010). They define nudges as default choices that do the least harm and most good, which should provide feedback for actions or inactions. A nudge should also align preferred behaviour, and simplify and facilitate decision making. The goals should in this sense also be visible, so that people can assess their performance towards a goal state. Similarly, Israeli-American author Nir Eyal states that human behaviour can be designed, but the main question to ask is if changing user's behaviour is advisable

(Eyal, 2014). This concerns the ethics of nudging; hence the critical question is if the solution will improve user's life, and how it can be valuable for them?

Design thinking can be characterized by a knowledge production that British science and technology scholar Michael Gibbons refers to as mode 2: problems that are solved in and for application (Gibbons, 1994). It seems that most designers and design researchers consider their work worthwhile, that it will make the user situation more valuable. The question is for whom? For a designer or a design researcher, such kind of knowledge production in practice often means increased interdisciplinarity and heterogeneity through collaboration with clients, users and other stakeholders. In such knowledge production, it is critical to know how to act in situations of complexity, uncertainty and value conflicts. According to previously mentioned American design theorist Donald Schön (1995), in contrast to the conventional lone designer/genius self-portrayal, it requires reflective practitioners who challenge their own perspectives. The sense-making that goes on in design thinking as well as in any other approach is grounded in the discourses that direct such practice. British science and technology scholar Lucy Suchman states that for the designers and/or design researchers, such situations include:

> "Overwhelmingly complex network of socio-material relations, for the most part made up of others (both human and non-human) we have never met and of which we were only dimly aware. /… / It is an increasingly dense and differentiated layering of people, activities and things, each operating within a limited sphere of knowing and acting that includes variously crude or sophisticated conceptualizations of others" (Suchman, 2002).

Partly as what can be seen as an answer to this quote, the Scandinavian participatory design (PD) tradition during the 1980s established an action-oriented practice of involving users in design (Bjerknes et al., 1987; Ehn, 2008), which now has emerged into an array of co-design practices (Sanders, 2002; Sanders and Stappers, 2008). Parallel to PD, French philosopher Bruno Latour developed actor-network-theory (ANT) as a way to deconstruct differences between actors, human as well as non-human, material or immaterial, in order to develop knowledge of how technologies, societies and cultures are interwoven (Latour, 1987). Among other things, this was seen as the responsibility to re-think design-after-design. ANT was in this sense a critique for thinking only of the design process, from problem to solution, and not taking responsibility for what happens after implementation and in use. British sociology scholar Judy Wajcman sees ANT as an opening to challenge misogynistic conceptions and patriarchal principles, and to intersect conceptual dichotomies such as mind vs body, active vs passive, objectivity vs subjectivity, and so forth, where the first idiom of each dichotomy is associated both to each other, and to masculinity and power (Wajcman, 1991).

One reason for such conceptions and principles can be found in semiotic and social constructions of reality. French semiotician Roland Barthes states that neither actors, nor things should be seen as neutral or transparent: meaning cannot be conveyed to us, we actively contribute in the sense-making activity (Barthes, 1972). Also, Swiss semiotician Ferdinand de Saussure describes the simplest form of communication, the symbol, as part of shaping social life, in that it consists of both the form/gestalt, and the representation/meaning that it takes (de Saussure, 2011). Barthes' concept of 'myths' explicates the storytelling our embedded sense-making activities build on within a particular culture. The function of myths is to create dominant historical and cultural norms, values, attitudes and ideas that are perceived as natural, normal, obvious and common-sense, that is, as objective and "true" reflections of how things are. British semiotician Daniel Chandler declares that we humans tend to create such stories for the things and situations that give privileges and power to ourselves (Chandler, 2007). It has also been said that designers have the ability to transform such myths into viable, solid and tangible forms, so that they appear to be "natural" and are perceived as reality itself (Forty, 1989).

British science and technology scholar Lucy Suchman argues that the relationship between humans and artefacts is interwoven, hence requires situated accountability (Suchman, 1987, 2002). This however involves challenging what American philosopher and feminist theorist Sandra Harding (1991) describes as the dichotomy between the researcher/designer/man as unemotional and neutral, and of the user/woman as emotional and biased, because it ignores how power, norms, values, myths, and metaphors play into the production of what is seen as valuable knowledge and who is seen as a relevant actor, in design as well as other practices. This poses a challenge of the designer's claimed neutrality, a standpoint which in this view only is beneficial for strong groups looking for foul play. It challenges the designers/knowers as objective and unbiased, and calls for accountability for reasons, intentions, and consequences. It involves a need for the most critical perspectives, to maximize value in the outcomes. As Harding argues, if women and African Americans are excluded, racist, sexist, and other norms and biases are not likely to be discovered by those who somehow benefit from such marginalization. This is also American science and feminist scholar Donna Haraway's motive, i.e., that there is a need for critical theory about how meaning and bodies are created, to be able to live in meaning and bodies that have a possible future (Haraway, 1988).

However, it might be challenging to develop a critical perspective, and virtually impossible to be in all critical positions of race, gender, class, age, and so forth, simultaneously. The situated production of knowledge that Haraway proposes in its place involves taking responsibility for nodes and directions, in material and semiotic meanings. This does not necessarily mean devoting all focus to different actors' pre-understandings, but taking responsibility in real life activities, challenging own and other's preunderstandings, espoused theories and self-images, and all actors' thinking of what they think they see (Haraway, 1997). As Haraway points out, there are always many actors involved in constructing knowledge, and there are always several possible stories. In a situated knowledge production, both designers and users should hence be seen as actors and agents in an exchange of knowledge, similar to what is proposed in co-design.

Design thinking can be seen as a social and cultural production of new visual, material and immaterial forms. In such production, as British human-computer interaction scholar Lucy Suchman states, no one should have neither control nor power, rather, the design process should be situated in human stories, weaves and connections, and action accountability (Suchman, 2002). Relevant in this aspect is also Swedish-American design scholar Erik Stolterman's emphasis on rejecting the neutrality or objectivity of the design method, as this would mean transferring knowledge and responsibility from the designer, to the method itself (Stolterman, 1991). The method should- and could -not be held responsible for providing guidance for actions, but perhaps it could support in nudging designers to think and act for human-centric accountability? Hence, designers could constantly be required to challenge and progress the methods, as well as the ideologies that guide them. The result of design should then not only be seen as new things, but as transformed ways of thinking, performing and acting among people, as well as new conventions and environments for design.

Knowledge production in design can also be seen as a form of practice. According to German sociology and philosophy scholar Andreas Reckwitz, practice theory does not focus only on individuals (micro) or organizations (macro), but on the nexus of thoughts, brains, bodies, things, knowledge, discourses, structures and processes and agency, i.e. the institutional arrangements within which design and its users are constituted (Reckwitz, 2002). To consider design as a practice involves making visible shared or collective symbolic knowledge structures, and the implicit and tacit norms that structure understandings of what is expected and seen as legitimate. Also, American feminist theorists Karen Barad states that practice can be understood as a configuration in which boundaries, properties, and meanings are differently enacted (Barad, 2007). A practice is a dynamic, local enablement through which multiple and different actors are interwoven into artful integrations. The material and the discursive practices are here interweaved into a dynamic intra-activity. This, Barad states, can be regarded as a phenomenon, produced by complex agential interactions of multiple, material-discursive practices or devices of physical production. The practice is in this view created in and through the socio-material world that it forms. Practices cannot be considered in isolation from their constitutional individual elements, but should be understood as a dynamic intra-play between different parts in relation to each other, as well as in relation to other subjects, social groups or networks.

French and British actor-network theorists Bruno Latour and Steve Woolgar highlight that practice can be seen as both global and local: so-called facts and reality are constructed based on larger worldviews, rather than being discovered on the spot or in the situation (Latour and Woolgar, 2013). There is thus a risk of representing and reproducing unequal power relations such as class, race, and gender (Asdal et al., 2001). The role for designers could relate to what Danish philosopher Søren Kierkegaard once said, in order to truly help someone else, one must understand more than that person, but certainly first and foremost understand what that person understands. If one does not do that, one's greater understanding does not help that person at all (Kierkegaard, 1859/1948). Adding to this could be a need to understand oneself and others in relation to power, privileges, and oppression as described in previous sections. In contrast to natural science, with its claims of ridding itself of politics and presuppositions, there is a need for design practice and design science to position itself through acknowledging the designers' and participants' standpoints, including norms, bias and presuppositions.

Thinking of acting in design

Design thinking generally involves acting, the previously described knowledge production of solving problems in and for application. In this, American design scholar Richard Buchanan recognizes that "we should consciously consider the possibility that our communications and constructions are, in some sense, forms of action. This does not deny the importance of information and physical embodiment, but makes us more sensitive to how human beings select and use products in daily life" (Buchanan, 1999). Buchanan continues saying that form, function, materials and manufacturing continues to be important in design thinking, but it is similarly important for design thinking to explore what makes a product useful, usable, and desirable. Likewise, American designer consultant Tim Brown and design scholar Barry Katz claim that the evolution of design into design thinking no longer limit design to "physical products, but also new sorts of processes, services, interactions, entertainment forms, and ways of communicating and collaborating"

(Brown and Katz, 2011). American design consultant and author Thomas Lockwood states that design thinking includes "applying a designer's sensibility and methods to problem solving, no matter what the problem is" (Lockwood, 2010), involving "a very clear shift towards a more creative and more collaborative way of working—one in which intuition counts heavily, experimentation happens fast, failures along the way are embraced as learning/.../and more relevant solutions are produced" (Lockwood, 2010). The question remains, what actions can be taken to realise if a solution is relevant and relevant for whom?

One answer to this can be the co-design approach, described by American design scholar Liz Sanders as a change of mindset. This mindset consists of considering all people as possible contributors to design, as long as the right tools are given for them to act. The rationale for a human-centred design approach is described as developing a deep understanding of the user's needs, desires and boundaries to meet these with design (e.g., Brown, 2008; Brown and Katz, 2011; Schneider and Stickdorn, 2011). Such a process is described in different ways, but with a common ground in being iterative and based in understanding the situation, empathizing with users and their context, as well as prototyping and testing in several steps to increase the understanding of what the solution needs to do. Austrian design scholar and consultant Marc Stickdorn believes that the difference in approach differs more in what methods are used than in the basic principles themselves: "Whatever you design, you must always understand the needs of users, you always work iteratively, you always have divergent and convergent phases" (Stickdorn et al., 2018). Regardless of which design thinking approach or method one employs, all phases are vital for the outcomes being a desirable, viable and feasible solution. However, what in some approaches are referred to as the understand or empathize phases, seems critical in terms of whose problem it identifies and explores, whose key concept the approach is being formulated upon, and what kind of thinking that is actually going on.

One implication of following steps and not thinking about what they include or not is however the idea of the user. Swedish design scholar Johan Redström states that the design subject used to be a matter of visual and material form, but now increasingly deals with the subject of the user and her experiences (Redström, 2006). In this sense, it should be understood "in the light of designs failing to get approval by users and situations where the intended use of designs does not translate into actual use—and how we as a design community have responded to this" (Redström, 2006). One implication of this is the Dutch and Irish design scholars Mike Robinson and Liam Bannon's conceptualization of ontological drifts (Robinson and Bannon, 1991). This is described as the translation of meaning that happens from the first collected user insights, passing through the filters of all of the different actors' and stakeholders' interpretations into a final implemented solution. In this, the process of interpretation and reinterpretation is central, as user's insights drift away from their original stated need, each actor interpreting and re-contextualising other's understandings into their own semantical, ontological, epistemological, and conventional belief systems. As stated: "if this is not taken into account in the design dialogues, the final outcome may come as a nasty surprise to all concerned. If it is taken into account, even in broad terms, expectations of the nature and results of the design process may change considerably" (Robinson and Bannon, 1991).

Another aspect to consider is the notion of tokenism. American advocate of participatory decision-making Sherry Arnstein in 1969 defined tokenism through a typology of eight levels of participation, ranging from (1) *non-participation* as manipulation, therapy, different degrees of (2) *tokenism* as informing, consultation, placation, to (3) *degrees of power* as partnership, delegated power, to control (Arnstein, 1969). In this, she stressed the critical difference between the empty ritual of user involvement and users having real power to affect the outcome of the process. Tokenism is in this sense defined as power-holders allowing the have-nots to hear and have a voice. American sociology and feminist theorist Rosabeth Moss Kanter similarly describes tokens as minorities in relation to dominating groups (Kanter, 1977). Tokenism in this sense deals with power, the negative consequences of inequality and blocked opportunities for minorities to contribute and have a say. In this sense, tokens are expected to act within their pre-defined minority role, generally women among men, but the dynamics of interaction is likely to be similar whether it is "blacks among whites, very old people among the young, straight people among gays, the blind among the sighted. /…/ Tokens are not merely deviants or people who differ from other group members along any one dimension. They are people identified by ascribed characteristics (master statues such as sex, race, religion, ethnic group, age, and so forth) or other characteristics that carry with them a set of assumptions about culture, status, and behaviour" (Kanter, 1977).

There are means to avoid user insights drift and tokenism in design. One such mean can be understanding situated practices. The dictionary description of 'situate' means to fix or build something in a certain place or position, to put something into context, or to be in a specified position.[1] The theoretical concept of 'situatedness' involves understanding that human experiences cannot be distinguished from environmental, social and cultural factors (Costello, 2014). American science and technology and feminist scholar Donna Haraway states that 'situated' involves

---

[1] Situate: https://www.lexico.com/en/definition/situate (2020-04-21).

understanding that both actors and actions need to be defined in terms of power, privileges, knowledge, norms, values and experiences in relation to a specific context (Haraway, 1988). American cognitive anthropologist Jean Lave and Swiss-American educational theorist Etienne Wenger describe that in a situated practice, the central elements are the participants/actors, the activities and the context (Lave and Wenger, 1991). In this lies an opportunity of re-creating design thinking practices and processes as situated and dependent on a lot of connections, connotations, and consequences for different people. The Finish-American architect Eliel Saarinen said that one should "always design a thing by considering it in its next larger context—a chair in a room, a room in a house, a house in an environment, an environment in a city plan" (Keller, 2014). A framework of situated design thinking hence needs to be nudging designer's into re-thinking own values and privileges, as well as re-thinking things as visual, material and immaterial qualities set in a larger system of thought that convey social and cultural representations and meanings.

There are several significances in the notions of users and actors, participation, user insight drifts, and tokenism, some being how to take accountability of user needs, how to avoid user's insights drifting away during the process, and how to deal with exceeding user's expectations through understanding their meaning *in situ*, rather than their espoused theories? Being in a minority position might mean that one cannot express all things explicitly, the actors hence need tools to express their meaning and desires through different means, besides having some say in the process and its outcomes to avoid their values and needs drifting away. There are several methods for exploring context and actors' knowledge and experiences, for example through exploring espoused theory and theory in use through observations (Schön, 1995), as retrospective analysis of both self and other's behaviours in different situations. Visual analysis can be a tool for identifying denoted meaning and connoted representations (Eriksson and Götlund, 2004). In this sense, visual representations contribute in signifying some things as natural, and there is a need for critical analyses of power, i.e., analysing the visual language that surrounds us every day as affective representations of how we see for example, women, young, old, people of different ethnic heritage, and people with different abilities. The spaces in between re-present and present are in this sense important, as they create room for cultural and ideological inscriptions. It is in these spaces that gender, ethnicity and 'normality' take its form, and through the efficient visual communication values are established and become the norm. Austrian communication scholar Florian Arendt describes neutralizing or thwarting the stereotypes in design and media as one way of reducing the effect of the stereotype itself, which can affect how we read and understand things in the future (Arendt, 2019). In line with this is some Swedish design scholars' articulation of norm-creative design including different tactics for nudging towards a transformation of norms and bias in design (Alves Silva et al., 2016; Wikberg Nilsson and Jahnke, 2018).

Dutch design scholar Froukje Sleeswijk Visser describes generative sessions such as workshops as means that through conscious planning and facilitation can bring attention to both implicit and explicit knowledge (Sleeswijk Visser, 2009). In this sense, interviews are ways of exploring people's espoused theory, how they think they would act; observations are means for analysing actor's embodied knowledge, participation in design activities can make actors describe both explicit and implicit knowledge about what they feel; and generative sessions such as workshops can through different means make actors explore their implicit theories through describing what they value and dream of. This can explicate norms, through both visual and other material that is used in the workshop. For example, developing personas and/or user scenarios and switching between different users in the stories are ways to identify norms, privileges and oppression (Källhammer and Wikberg Nilsson, 2012). Critical incident technique is a method for assembling human behaviour, to identify incidents that have been critical, either positively or negatively (Flanagan, 1954). Using this method for identifying privileges and oppression in routine behaviours might make people more aware of how it affects or influences self and others, what role different actors play in a specific scenario, either by contributing or by not acting at all. Other ways are gender observations, gender system analysis and world café as methods that supports critical explorations and reflections of what is (Fältholm et al., 2016). Methods that aim for understanding both own and other's values and belief-systems can be tools in workshops or other sessions, for example 'value exercises', 'get the point', 'broaden the horizon', 'rewind the tape', 'provotypes' as in provoking examples, or 'roleplays' (Alves Silva et al., 2016).

Towards a framework of situated design thinking

Design practice can be said to include creating and changing society through processes that are at the same time intentional, situated and emerging. A design process should neither focus solely on the user, nor on the design task as such, but needs to take in the entire system in which the final design should interrelate. One way of thinking of design is through American design scholar Richard Buchanan's four orders, which are intertwined and contingent to one another: *symbols*, as things talking to a person; *objects*, as things a person interacts with; *interactions*, as groups of people and things in interaction; and *systems*, as groups of people and things in interaction with other groups of people and things (Buchanan, 1992). These can be seen as independent design practices or professions, e.g., graphic design/visual design, industrial design/architecture, service design/interaction design/UX design, and systems design/business design. However, they can also be seen as elements that are interdependent fragments of a larger whole. As such, they can be used as a framework for sense-making, storytelling and re-construction of norms in design, i.e., we see, do and reflect on certain things and not others. Design is not merely a practice of making things, it is a practice of transforming views of subject matter held by designers and others, and the material and immaterial things that are envisioned, planned and produced are expressions of those views (Buchanan, 1992).

The four orders of design should in this sense be seen as places of invention, in which one might discover different connections, connotations, and consequences of design by a situated design re-thinking of every design aspect in relation to the next, as: "signs and images are fragments of experience that reflect our perception of material objects. Material objects, in turn, become instruments of action. Signs, things, and actions are organized in complex environments by a unifying idea or thought [i.e., a system]" (Buchanan, 1992). Situated design thinking can be one concrete way of interplay and interconnection of signs, things, actions, and thoughts, through posing questions to explore what participating actors—whether designers, users or other stakeholders—see, do and reflect upon. The aim of nudging actors to more valuable and norm-creative design actions through these four lenses can be seen in Figure 2.

A framework of situated design thinking needs to address real situations, designers and users are part of, and challenge unconscious and uninformed routine thinking of humans and artefacts, integrating co-explorations as learning new with others, which is situated in different people's situations and experiences. The core of such a framework is to be nudging actors, both designers and other participants and stakeholders, into thinking through the situation *in situ*, i.e., on site and in position, as in realizing different stories, values, privileges, and oppressions, c.f. Figure 1. The idea is that this stimulates system 2 thinking through realizing that knowledge has to build on both own understandings and others', and through human-centric accountability. The lenses in Figure 2 do not illustrate the norm-critical phase of challenging what is, and the norm-co-creative phase of exploring what ought to be. This can be implemented through different means, for example the previously mentioned methods and techniques, through mixing new kind of visual, material, social and

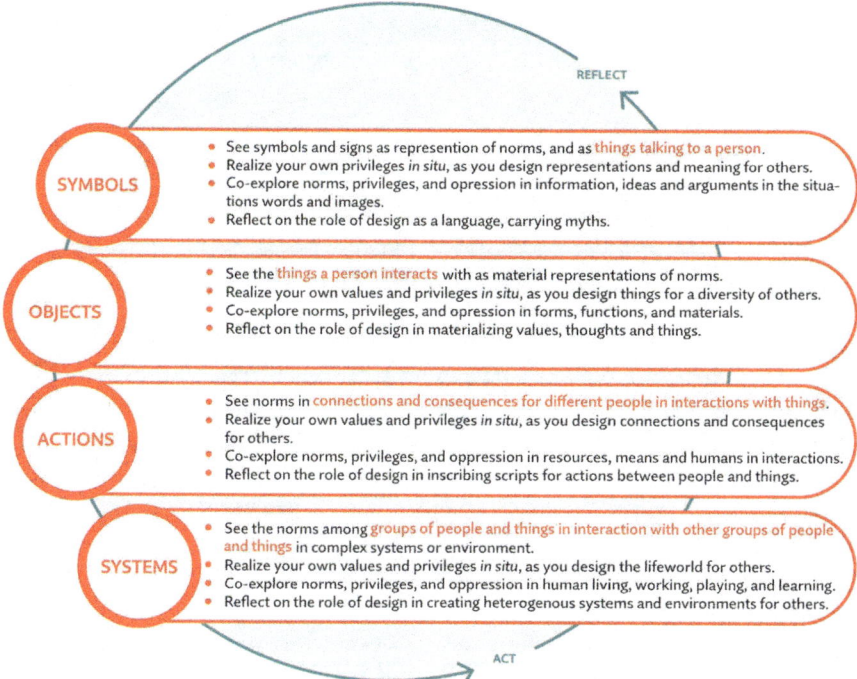

Figure 2. Illustrating lenses for nudging into situated design thinking. Illustration: Åsa Wikberg Nilsson.

cultural relationships between people and things, and creating intelligent, meaningful, and satisfying interactions for a diversity of others in the new landscapes of design. That accountability should lie on the designer/ design researcher. Guidance for this can be a situated design thinking model, see Figure 3, which builds on the intersectional axes in Figure 1 and the lenses in Figure 2 and together with them form the grounds of a framework for situated design thinking.

The situated design thinking model involves the phases of empathize, co-define, co-create, co-explore, co-apply and follow-up. The intention behind these phases is to highlight the need for including a diversity of bodies and voices in the design process, hence the indicative co- in front of four of the phases. The first phase of this model indicates the need for realizing own and other actors' norms, privileges, and oppressions *in situ*, in context, and the final phase indicates the importance of continuous learning through following-up what actually happens to designs after design.

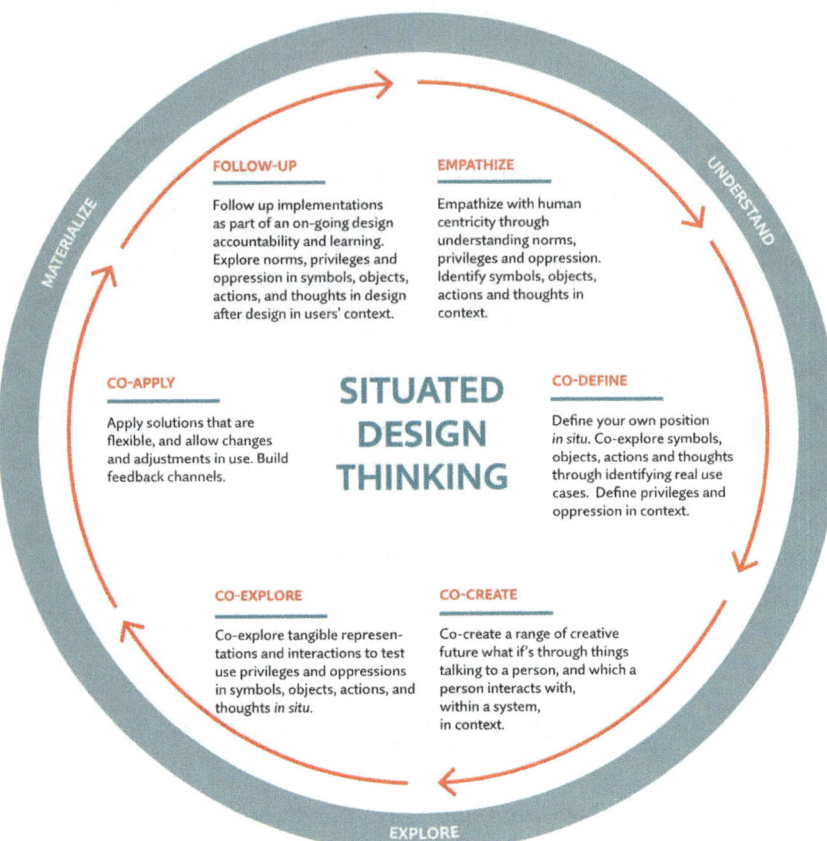

Figure 3. Illustrates a model for Situated Design thinking. Illustration: Åsa Wikberg Nilsson.

Conclusion

In this text, I have been exploring grounds for situated design thinking. In short, this can be explained from my standpoint of design needing a situational awareness, i.e., to critically explore and understand designers, users and design(s) representations and meaning, *in situ*, and through both bodies and minds. Design as a situated practice might involve a change of mindset for some, as Suchman (2002) says: it challenges the prevailing myth of the sole creator (the designer) and the passive recipient of design (the user). The reason for this standpoint is that an outlook on the world should be represented as a view from somewhere, a challenge of actors' beliefs, prior understandings, and judgmental heuristics and biases, which is always an embodied and partial perspective that we as designers and design researchers should be accountable for. As has been shown, design is not norm-neutral, it is a practice that is made up of people and that makes things for people, and the norms that guide people's un-reflected thinking and actions constitute both the practice and its outcomes. The me-too movement, demonstrating against sexual harassments and abuse initiated in the autumn of 2017, and the demonstrations by the black lives matter movement during 2020, indicate that some people are fed up of marginalization and oppression. This cannot be limited to a women's or black lives' issue, it concerns all people. Design has the ability to make such experiences count through retrospective and forethoughtful analysis of values, practices, and cultures, and outcomes.

On a *micro*-level, this can be stated as a need for designers and other actors in a design process to outline their own standpoints in relation to the specific design situation, and strive to involve a diversity of actors in co-explorations of tangible and intangible *in situ* symbols, objects, actions, and systems. On a *meso*-level, it might be stated as a need for design practices exploring how norms, privileges, and oppressions are embedded in cultures and ideologies, and how this is represented in design methods, -means, and -outcomes. Co-design might for that reason represent a re-thinking of human-centricity in terms of inclusion and diversity, given that actors develop empathy with power, privileges and oppressions. On a *macro*-level, it affects policies as well as design principles and practices in organisations and society as a whole, and a need to explore what is valued as viable, feasible, and desirable design(s) in the everyday things, everyday people face, every day. Make them count.

The following sections summarize a design rationale for situated design thinking. *Firstly*, the main idea of situated design thinking is that design(s), contribute in reconstructing and upholding norms and biases in social and cultural contexts. Just as in any other research or practice area, design thinkers need to re-think both explicit and implicit values, privileges, representations, connections, and consequences involved in designerly knowing, thinking, and acting. Design thinkers need to draw upon some of its traditions and experiences, whilst building a genuine human-centric intellectual and situated design thinking culture. This is not a "quick fix", but might involve genuine accountability for human-centricity and empathy, and transforming design thinking "towards new integrations of signs, things, actions, and environments that address the concrete needs and values of human beings in diverse circumstances" (Buchanan, 1992).

Secondly, the phases of developing understanding and empathy, in some of the design-thinking approaches, doesn't by default connect to human-centricity, insofar as it does neither require agreement from regarded or disregarded users, nor stating which views of things and others designers and other actors have. The central design thinking skill therefore lies in developing a sense of "speaking design" in a way a different actor can value, and the ability of reflective thinking as a way to increase awareness and sensitivity (Löwgren and Stolterman, 1998). Learning by doing seems often promoted in some design practices, however, as American pragmatists McLellan and Dewey (2007) stress, it involves learn to do by knowing, and to learn by doing, hence reflection on how we think and what we think is vital. An advice can be to be aware of one of the basic patterns in learning-by doing, i.e., that you follow someone without knowing what you are learning or why you are doing the things you do. Instead, new designers/learners should explore how concepts, values, and other things-that-counts are co- and re-constructed at different times and places (Haraway, 1988).

Thirdly, societies, cultures and various social groups act and are influenced by meta-stories that govern our way of perceiving the world. By merely copying the existing stories and possibly adding an African-American hero or a woman, there is a risk that the unequal power relations and stereotypical social norms are repeated, rather than transformed. The stories we humans use to make sense of things are always individual and particular, but at the same time refer to overarching global sensemaking, such as norms or myths, which provide certain values and perspectives as "right" or "normal". The stories people tell are reflections and projections of human actions and experiences, they are sometimes liberating and constructive, but sometimes oppressive and destructive. For the sake of the first design principle of human dignity and human rights (Buchanan, 2001), the designer should have accountability for creating the lifeworld for a diversity of people.

Fourthly, the different symbols, things, actions, and systems that are considered in design thinking ought to be valued on the basis of critical inclusion and evaluation of power balances and norms, and include personal responsibility and risk taking, as well as a radical humility and empathy for diverse user experiences, at least to the same account as aesthetic, economic, and technical aspects are valued today. We are designing as people, and we are designing for people. For this reason, who is telling the stories or creating the knowledge is a crucial question that needs always to be asked and answered to be able to deliver value to the people in the situation at hand.

Finally, some concluding thoughts. Following a situated design thinking framework does not guarantee human-centred design(s). It does however allow for a critical rethinking of existing discourses and the technological, economic, cultural, and social incentives formulated within them. It assures conscious actions taken applying the system-2 situated design thinking through continuously seeing, doing and reflecting on design through different lenses in concrete situations. Situated design thinking, in a paraphrase to Lockwood (2010), includes nudging designers through situated awareness to transform social environments in which people are able to live, love, and laugh. It involves a clear shift towards more conscious ways of doing design, one in which reflective thinking counts heavily, co-explorations happen fast and with a lot of voices and bodies, failures along the way are embraced as new learnings, and in which designers consider themselves as deeply dedicated visitors in the lifeworld of everyday people. For a true neoteric art of design, there is a need for situated design thinking as one step to counter sublime things taken-for-granted, and thus enable a genuine human-centric 'we-methodology' of co-exploration, co-creation and co-responsibility, where new insights can be achieved and new design(s) can be cultivated.

References

Akrich, M. 1992. The description of technical objects. pp. 205–224. *In*: Bijker and Law's [eds.]. Shaping Technology-Building Society: Studies in Sociotechnical Change. MIT Press, Cambridge.

Akrich, M. 1995. User representations: practices, methods, and sociology. pp. 167–184. *In*: Rip, A., Misa, T.J. and Schot, J.'s [eds.]. Managing Technology in Society. The Approach of Constructive Technology, Assessment Pinter Publisher, London/New York.

Alves Silva, M., Ehrnberger, K., Jahnke, M. and Wikberg Nilsson, Å. 2016. NOVA- Methods and Tools for Norm-creative Innovation, Vinnova, Stockholm.

Appleton-Dyer, S. and Field, A. 2014. Understanding the Factors that Contribute to Social Exclusion of Disabled People: Rapid Review for Think, Ministry of Social Development, Auckland, NZ.

Arendt, F. 2019. Investigating the negation of media stereotypes. Journal of Media Psychology, 31: 48–54.

Argyris, C. and Schön, D. 1975. Theory in Practice: Increasing Professional Effectiveness. Jossey-Bass, San Fransisco.

Arnstein, S.R. 1969. A ladder of citizen participation. Journal of the American Institute of Planners, 35(4): 216–224.

Asdal, K., Brenna, B. and Moser, I. 2001. Teknovitenskaplige kulturer. Spartacus Forlag, Oslo.

Barad, K. 2007. Meeting the universe halfway: quantum physics and the entanglement of matter and meaning. Duke University Press, Durham, NC.

Barthes, R. 1972. Mythologies. The Noonday Press, New York.

Barthes, R. 1993. Rhetoric of the image. pp. 152–163. *In*: Heath's, S. [ed.]. Image, Music, Text: Essays Selected and Translated by Stephen Heath. Macmillan, New York.

Berg, A.-J. and Lie, M. 1995. Feminism and constructivism: do artefacts have gender? Science, Technology & Human Values, 20(3): 332–351.

Berg, J. 2010. Race, class, gender, and social space: using an intersectional approach to study immigration attitudes. The Sociological Quartely, 51(2): 278–302.

Bjerknes, G., Ehn, P. and Kyng, M. 1987. Computers and Democracy. Aldershot, Brookfield, US.

Brown, T. 2008. Design Thinking. Harvard Business Review, Volume June: 84–92.

Brown, T. and Katz, B. 2011. Change by design. The Journal of Product Innovation Management, 28(3): 381–383.

Buchanan, R. 1992. Wicked problems in design thinking. Design Issues, 7(2): 5–21.

Buchanan, R. 1999. Design research and the new learning. Design Issues, 17(4): 1–22.
Buchanan, R. 2001. Human dignity and human rights: thoughts on the principles of human-centered design. Design Issues, 17(3): 35–39.
Butler, J. 2011. Gender Trouble: Feminism and the Subversion of Identity. Routledge, Abingdon, UK.
Chandler, D. 2007. Semiotics: The Basics. Taylor & Francis, Abingdon, UK.
Cockburn, C. and Ormrod, S. 1993. Gender and Technology in the Making. Sage Publications Ltd, London, UK.
Costello, M. 2014. Situatedness. pp. 1757–1762. In: Teo, T. [eds.]. Encyclopedia of Critical Psychology. Springer, New York.
Crenshaw, K. 1989. Demarginalizing the intersection of race and sex: A black feminist critique of antidiscimination doctrine, feminist theory and antiracist politics. University of Chicago Legal Forum, pp. 139–167.
de Beauvoir, S. 1997. The Second Sex. Vintage, London.
de Saussure, F. 2011. Course in General Linguistics. Columbia University Press, New York.
Dilnot, C. 1982. Design as a socially significant activity: an introduction. Design Studies, 3(3): 139–146.
Dorst, K. 2008. Design research: a revolution-waiting-to-happen. Design Studies, 29(2008): 4–11.
Ehn, P. 2008. Participation in Design Things. Indiana University, Bloomington, pp. 92–101.
Ehrnberger, K., Räsänen, M. and Ilstedt, S. 2012. Visualing gender norms in design: Meet the Mega Hurricane Mixer and the Drill Dolphin. International Journal of Design, 6(3): 85–98.
Eriksson, Y. and Götlund, A. 2004. Möten med bilder. Studentlitteratur, Lund.
Eyal, N. 2014. Hooked: How to Build Habit-Forming Products. Penguin, London.
Fältholm, Y., Wennberg, P. and Wikberg Nilsson, Å. 2016. Promoting Sustainable Change: A toolkit for integrating gender equality and diversity in research and innovation systems. [Online] Available at: https://genovate.cdt.ltu.se/.
Fiske, S.T. and Taylor, S.E. 1991. Social cognition. McGraw-Hill, New York.
Flanagan, J.C. 1954. The critical incident technique. Psychological Bulletin, Volume July (1954): 327–358.
Forty, A.J. 1989. Objects of desire. University of London, London.
Foucault, M. 1990. The History of Sexuality: An Introduction, Volume I. Vintage, New York.
Friberg, T. and Larsson, A. 1999. Om kvinnligt och manligt i planeringens könsneutrala värld. Nordisk arkitekturforskning, 2: 33–44.
Gibbons, M. 1994. The new production of knowledge. The Dynamics of Science and Research in Contemporary Societies. Sage, London.
Gibbs, J.P. 1989. Control: Sociology's Central Notion. University of Illinois Press, Urbana, Ill.
Haraway, D. 1988. Situated knowledge: the science question in feminism and the privilege of partial perspective. Feminist Studies, 14(3): 575–599.
Haraway, D. 1997. Modest Witness@Second Millenium. FemaleMan meets OncoMouse. Feminisim and TechnoScience. Routledge, New York.
Harding, S. 1991. Whose Science? Whose Knowledge? Thinking from women's lives. Open University Press, Buckingham.
Julier, G. 2005. From visual culture to design culture. Design Issues, 22(1): 6476.
Julier, G. 2008. The Culture of Design. Sage, London.
Kahneman, D. 2013. Thinking fast and slow (Tänka snabbt och långsamt in swedish). Månpocket, Stockholm.
Källhammer, E. and Wikberg Nilsson, Å. 2012. Gendered innovative design—critical reflections stimulated by personas. pp. 328–350. In: Andersson, S., Berglund, K., Gunnarsson, E. and Sundin's, E. [eds.]. Promoting Innovation: Policies, Practices and Procedures, Gendering Innovation. Vinnova, Stockholm.
Kanter, R. 1977. Some effects on proportions on group life: Skewed sex ratios and responses to token women. American Journal of Sociology, 82(5): 965–990.

Keller, H. 2014. AD remembers the extraordinary work of Eliel and Eero Saarinen. [Online]: https://www.architecturaldigest.com/story/saarinen-father-and-son.
Kierkegaard, S.A. 1859/1948. Synspunktet for min forfatter-virksomhed (The viewpoint for my authorship): l. Finn Wangberg.
Kirkham, P. 1996. The gendered object. Manchester University Press, Manchester.
Löwgren, J. and Stolterman, E. 1998. Design av informationstekik - materialet utan egenskaper. Studentlitteratur, Lund.
Latour, B. 1987. Science in Action. How to follow Scientist and Engineers through Society. Harvard University Press, Cambridge, MA.
Latour, B. and Woolgar, S. 2013. Laboratory of Life: The Construction of Scientific Facts. Princeton University Press, Princeton, NJ.
Lave, J. and Wenger, E. 1991. Situated Learning: Legitimate Peripheral Participation. Cambridge University Press, Cambridge.
Lee, N. 2012. Culturaly responsive teaching for 21st century art education: Examining race in a studio art experience. Art Education, 65(5): 48–53.
Lidwell, W., Holden, K. and Butler, J. 2010. Universal Principles of Design: 125 Ways to enchance usability, influence perception, increase appeal, make better design decisions, and teach through design. Rockport Publishers, Beverly, MA.
Lockwood, T. 2010. Design Thinking. Integrating Innovation, Customer Experience, and Brand Value. Simon and Schuster.
McLellan, J.A. and Dewey, J. 2007. Applied Psychology. An introduction to the principles and practice of education. Educational Publishing company, Boston.
Molotoch, H. 2003. Where Stuff Comes From: How Toasters, Toilets, Cars, Computers and Many Other Things Come to be as they are. Routledge, London.
Morewedge, C.K. and Kahneman, D. 2010. Associative processes in intuitive judgment. Trends in Cognitive Sciences, 14(10): 435–440.
Morgan, K.P. 1996. Describing the Emperor's New Clothes: Three myths of educational (in) equity. pp. 105–122. In: Diller, A. et al. [eds.]. The Gender Question in Education— Theory, Pedagogy, and Politics. Routledge, New York.
Navarro, R., Martinez, V., Yubero, S. and Larranaga, E. 2014. Impact of gender and the stereotyped nature of illustrations on choice of color: replica of the study by Karniol (2011) in a Spanish sample. Gender Issues, 31(2): 142–162.
Norman, D.A. 2003. Emotional Design: Why we Love or Hate Everyday Things. Basic books, New York.
Norman, D.A., Ortony, A. and Russel, D.M. 2003. Affect and machine design: lessons for the development of autonomuous machines. IBM Systems Journal, 42(1): 38–44.
Papanek, V. and Buckminster Fuller, R. 1972. Design for the real world. Thames and Hudson, London.
Peterson McIntyre, M. 2013. Perfume packaging, seducation and gender. culture unbound. Journal of Current Cultural Research, 5: 291–311.
Peterson McIntyre, M. 2018. Gender by design: Performativity and consumer packaging. Design and Culture, the Journal of Design Studies Forum, 10(3): 337–358.
Pile, J.F. 1979. Design: Purpose, form, and Meaning. University of Massachusetts Press, Amherst.
Reckwitz, A. 2002. Towards a theory of social practices. A development in Culturalist Theorizing. European Journal of Social Theory, 5(2): 43–263.
Redström, J. 2006. Towards user design? On the shift from object to users as subject of design. Design Studies, 27(2): 123–139.
Reimer, S. 2016. Its just a very male industry: gender and work in UK design agencies. Gender, Place & Culture, 23(7): 1033–1046.
Rittel, H.W.J. and Webber, M.M. 1973. Dilemmas in general theory of planning. Policy Sciences, 4(1973): 155–169.

Robinson, M. and Bannon, L. 1991. Questioning Representations. Second European Conference on Computer-Supported Cooperative Work, Amsterdam, the Netherlands, pp. 219–233.

Ruble, T.L. 1983. Sex stereotypes: issues of change in the 1970s. Sex Roles, 9: 397–402.

Sanders, E.B.-N. 2002. From user-centered to participatory design approaches. pp. 18–25. *In*: Frascara, J. [ed.]. Design and the Social Sciences. Taylor & Francis Books Limited.

Sanders, E.B.-N. and Stappers, P.J. 2008. Co-creation and the new landscapes of design. Co-design, 4(1): 5–18.

Schön, D. 1985. The Design Studio: An Exploration of its Traditions and Potentials. RIBA, London.

Schön, D. 1995. The Reflective Practitioner: How Professionals think in Action. Arena, Aldershot.

Schneider, J. and Stickdorn, M. 2011. This is Service Design thinking: Basics, Tools, Cases. Wiley, Hoboken, NJ.

Sleeswijk Visser, F. 2009. Bringing the everyday life of people into design. Se Niewe Grafische, Rotterdam.

Sparke, P. 2010. As Long as it's Pink: The Sexual Politics of Taste. Press of the Nova Scotia College of Art and Design, Halifax, N.S.

Star, S.L. 1991. The sociology of the invisible: the primacy of work in the writings of Anselm Strauss. pp. 265–283. *In*: Maines, D. [ed.]. Social Organization and Social Process: Essays in Honor of Anselm Strauss. Aldine de Gruyter, New York.

Stickdorn, M., Horness, M., Lawrence, A. and Schneider, J. 2018. This is Service Design Doing: Applying Service Design thinking in the Real World. O'Reilly Media Inc, Sebastopol, CA.

Stolterman, E. 1991. Designarbetets dolda rationalitet. Umeå University, Umeå.

Suchman, L. 1987. Plans and Situated Actions. The Problem of Human-machine Communication. Cambridge University Press, New York.

Suchman, L. 2002. Located accountabilities in technology production. Scandinavian Journal of Information Systems, 14(2): 7.

Thaler, R. and Sunstein, C. 2009. Nudge: Improving Decisions about Health, Wealth, and Happiness. Penguin, London.

Tversky, A. and Kahneman, D. 1974. Judgment and uncertainty: Heuristics and biases. Science, 185(4157): 1124–1131.

Wajcman, J. 1991. Feminism Confronts Technology. Polity Press, Cambridge.

West, C. and Zimmerman, D.H. 1987. Doing gender. Gender and Society, 1(2): 125–151.

Wikberg Nilsson, Å. and Jahnke, M. 2018. Tactics for norm-creative innovation. She-Ji The International Journal of Design, Economics, and Innovation, 4(4): 375–391.

CHAPTER-5

The Challenge of Designing Meaningful Information

*Yvonne Eriksson** and *Anna-Lena Carlsson*

Information design is a relatively new design discipline. In 2000, Robert E. Horn defined information design as "the art and science of preparing information so that it can be used by human beings with efficiency and effectiveness" (Horn, 2000). The objectives behind this statement were to develop documents that are comprehensible, to design interfaces that allow interactions that are coherent and easy to use, and to enable people to find their way in different spaces (ibid.). So far in its short history, information design has been focused on how to organize information of various kinds, such as in public spaces that contain different signs, and governmental information or information about healthcare services. This goes in line with the common understanding of the concept of *information* as an ordered reality, and something that could be distributed (Dervin, 2000) which implies that information is regarded as an artefact, that is, something that exists over time with a fixed meaning. Taking a socio-cultural perspective on information, it can be defined as something that changes over time in relation to the context it appears within and the situation in which it is perceived. This is because the way information is used and understood depends on the context, the situation, and people's motivation for using it. If we regard information as something that is not static, even though the information artefact itself is unchanged, the understanding and interpretation will be modified over time and understood differently between individuals and groups. How individuals and groups perceive, understand, and use information depends on aspects such as socio-economic and cultural background, gender, age, ethnicity,

---

School of innovation, design and engineering. Division of Information Design, Eskilstuna/ Västerås, Mälardalen University.
* Corresponding author: yvonne.eriksson@mdh.se

education level, cognitive and perceptual ability, and willingness to perceive and/or motivation to use the actual information. Therefore, it is relevant to talk about information literacy, even though the perception of information varies.

In this chapter, we aim to problematize information as something that is built on signs and risks ambiguity, depending on the context, situation, and the perceiver, where perception is viewed as an active process. Furthermore, we argue that the perception of meaning, and the meaning-making of information is crucial for its impact. So, the challenge for information designers and information design research is to frame *what* should be communicated, *why*, and *how* the information should be designed in order communicate it in a way that will be perceived as meaningful.

Different methods and perspectives

Much of the research related to information design takes place in different areas and is not always defined as information design. This could be explained by the subjects' short history, and by the written history's often partial character due to the author's residence, as well as by information design's interdisciplinary character. Research within cognitive science, behavior psychology, linguistics, visual studies, and social science, for instance, are relevant for information design. However, research does exist in the name of information design, and several methods are used in information design research, most of which are qualitative and take an ethnographic approach. Interviews, observations, artefact analyses, and photo-elicitations are frequently used in combination with approaches such as research through design, participatory design, situated design, and design thinking. Research through design is understood as explorative research, where the starting point is a need for change or a new product. Step by step, the information artefact or services are developed in an iterative process where individuals from the intended target groups are involved in a continuous cycle of evaluation.

In participatory design, the process focuses on the involvement of the participants to a greater extent than in research through design (Di Salvo et al., 2017). Situated design research is located in the actual context of the change needed. If universal methods of design (Martin and Hanington, 2012) are aimed at designing for the many, situated design is embedded in the situation and participants related to the design. Situated design research, is also a form of research through design (Baerenholt et al., 2014; Simonsen et al., 2014). Also, knowledge is produced within its situation of application and, accordingly, validation takes place through its use. However, paying attention to both the universality and the situatedness of a design suits information design particularly well. This is because communication from, for example, authorities during a national crisis is dependent on it functioning in both a situated and wider context. It speaks to a whole population and also needs to be understood in particular situations. Moreover, the particular need of information design changes. For example, in an assembly in a multinational manufacturing company, information design may need to meet the needs of both the literacy of the local operators and the needs of a global standard (Eriksson et al., 2011).

Design Thinking is used in information design research, and student projects, and its heritage can be traced to the IDEO studio in Palo Alto (Brown, 2009). Design Thinking takes a *human centered design* perspective. Both methods were developed in the Design Thinking context, and the definition of Design Thinking is used and interpreted in various ways. It is used as a business model for innovation, but it is also used in design research. In design research, Design Thinking follows several steps: *(1) Empathize with the user, (2) Define the problem, (3) Ideation, (4) Prototyping and testing.* This process can be carried out mechanically. However, with careful use of the process involving both external partners (such as co-designers) and end users, it is possible to develop what is often referred to as *usable information design* built out of signs with letters, images, among others, as well as physical artefacts of different kinds where information is embedded. Information design aims to fulfill its intended purpose. However, what does it mean to adopt a "human centered" design perspective? To what extent is it possible to talk about usable information, since its usability is dependent on how it is perceived, interpreted, and understood? The universal approach, perhaps with support from practices or research within Plain or Easy-to-read language, has situatedness in focus, with a work-specific or language-in-use focus. Both approaches, used separately or combined if necessary and possible, pay attention to the insight or change related to the communication.

In a research context, when information designers involve the expected target group and/or individual users of the information, it also gives them the opportunity to identify themselves with various groups of people and their living conditions. This can help address questions such as: What does it mean to be a woman or man, old or young, have a disability, limited language skills, be unemployed, or just be in a particular situation? These, and many other aspects of life, have a huge impact on how one perceives, interprets, and uses information. Information design offers the opportunity for emotional involvement, or for gaining a better understanding of how people behave or use information in specific situations. It also offers the ability to design for specific groups. However, when designing for specific groups in a research context, there is a tendency to define users by applying simplified categorization and classification schemas. Such categorization often leads to stereotypic descriptions of individuals and groups of people, which can hinder the ability to gain a deeper understanding of people's needs regarding specific information or interfaces (Eriksson, 2016). Therefore, it is important to reflect upon the aim of the information and what should be communicated. What might be obvious to one person may not necessarily be relevant for another. In this sense, from the perspective of a specific group of people, information could even be regarded as something that should be ignored. Or the act of ignoring societal information could become a political statement, if the information is experienced as state abuse. As an information designer, it is particularly crucial to recognize that we must consider both standardization and the living.

An information designer needs to know why a given message is posted and, especially, to understand the intention behind it. It starts with the mental model of the sender, and the aim of the sender's message is to try to influence the mental model of the recipient. To be able to do so, it is necessary to get clear on what we want to achieve. Is our aim to guide people to the closest fire exit in a specific building, or to persuade them to stop smoking or to start exercising? The guide to the fire exit requires that we are able to put the sign in a location that makes it visible, and that we use a conventional format that people can recognize. Meanwhile, it is more challenging to persuade people to change their behavior, for instance, to stop smoking or start exercising. This can be explained by the fact that, in order to use information, most human beings need to feel that they are being spoken to. It is not enough to present the negative effects of smoking and the different risks involved, such as cancer and cardiovascular diseases, since such repercussions belong to the future and it may be difficult to not only grasp, but to imagine themselves being affected by these kinds of diseases. Also, information might not be the problem. The acquired physical need overrides the information, especially when the consequences seem far away.

Therefore, such information not only encounters a lot of challenges, but is also a hindrance. Socio-semiotic structures within a group are related to the mental image of the members. This affects habits and behaviors, as well as acquired physical needs among individuals, and is often much stronger than information that comes from outside especially when it comes from authorities (Chimuanya and Ajiboye, 2016). In addition, the group's behavior and interpretation influences the mental images among individuals. No matter how well it is designed, information will be useless if it is not used or understood. It is necessary to find a way to break through the barrier of people's assumptions and behaviors, as well as the socio-semiotic structures of groups, in order to reach the aim of a specific piece of information. For instance, it took many decades before people recognized the need for an environmentally sustainable lifestyle (and still, many people fail to recognize this). It wasn't until people witnessed the consequences of climate change that they started to pay attention to information concerning knowledge on sustainability issues. However, this is assuming that the information contains a message based on scientific studies and, thereby, is presented as fact. However, to some extent, fact is a relative concept and sometimes humans only listen to what they want to hear or disregard fact, knowledge, and information, because it suits other purposes they have. How a fact is interpreted is not only a question of how the recipient interprets it, but it is also about the information designer's interpretation. Designed information based on research is multi-layered. It requires an interdisciplinary approach, where cognitive, behavioral, and cultural aspects are taken into consideration, as well as how the information should be designed in order to be effective and efficient. What should be communicated, why, and how can the aim of the communication act be reached?

Communication: from perceiving information to sharing with others

To use information means to internalize a message and make it one's own, i.e., incorporate it into one's own understanding and grounds for action. This leads to the relationship between information and communication. As mentioned before, information design is about communication. The etymological explanation of communication comes from the Latin word *communicare*, which means to have in common, to share the same understanding of something. The Swedish scholar Peder Hård af Segerstad (2009) defines the communication process as follows: *Knowledge of the carrier—information transmitted—the information interacts with the recipient's information system (the mental image)—new information is created—knowledge of the recipient—the recipient spreads the information further.* From the discussion in the previous section, we can conclude that communicating information is a complex process. However, the process recommended offers a structure that is neither mechanical nor ignores the directional nature of information in communication (someone is *informing* someone else). If it also takes a design thinking perspective into consideration, and each stage contains an iterative process—especially if the information designer takes the process seriously.

From an information design perspective, we can try to follow the defined process of communication by Hård af Segerstad in combination with the Design Thinking model. In this case, the first step for an information designer would be to determine what is most important to communicate, what the message is and what its desired effect is. It is important to reflect upon the expected outcome of the information and frame the challenge or the problem. The next step is to empathize with the individuals one intends to reach. To do so means to gain an understanding of a specific group's living conditions, or diversity in living conditions and socio-economic standards; or to gain an understanding of how a larger group of people, such as citizens in a region or country, lives. Some designers use an anthropological method to gain an understanding of the living conditions for people with disabilities (Gunn and Donovan, 2012). They use wheelchairs, put on glasses that reduced their eyesight, and went out in the city every day for a period of time. Here, they use role play as a method for getting a bodily insight. Through that experience, they gain a fuller understanding of how vulnerable a person with disabilities can be. This may be regarded as an extreme method but, still, it is important to try to walk in someone else's shoes. Doing so might lead to a re-definition of the original idea about how the information should be designed. Working with an early prototype of the information artefact, it is possible to test, evaluate and re-design it until it fulfills expectations. Once the information is launched, it is possible to investigate how the information is perceived and interacted with by the recipient, and what kinds of new information is generated. Has the information design produced the expected effect and impact, or not? If not, the content of the information might be wrongly formulated or a change in focus could be needed. But the media or the design of the media could also be to blame for the failure. Very often, it is argued that people lack information literacy, or that it is of poor quality since it does not achieve its aim. In taking this process seriously, it becomes obvious that it takes time, and many iterations are required before the target is reached.

The relationship between the interpretation of information and situation awareness

In 1964, Marshall McLuhan phrased that the medium is embedded in the message, since there is a symbiotic relationship between how the message is perceived and the medium that has been used (McLuhan, 1964). Designed information is not only embedded in a text, visualization or space, but it is also embedded in the medium as such, and in the design itself. Embedded information is not explicit, it affects the perception indirectly. In a well-designed, permanent way showing system, information is embedded so that it lasts over time (see Figure 1) while information embedded in a temporary way showing system, as in Figure 2, is an ad hoc solution.

How will the way we perceive information be impacted if the information is displayed on a computer screen, poster, or in a newspaper? Or how will our perception be impacted if information is primarily made up of visuals versus texts? The conditions for how we perceive and interpret different media and its content depends on the design of the visual or text. A text presented in a linear form will be viewed differently than the same text presented in a series of bullet points. Likewise, a visual narration presented on a single sheet will be read differently if it is divided

Figure 1. Information that last over time. Photo: Yvonne Eriksson.

Figure 2. Information that is temporary. Photo: Yvonne Eriksson.

into the frames of a comic strip. Graphics, formatting, and hierarchical structures also play a role in the meaning-making process.

Visual media is a broad concept. It includes various kinds of pictures (sketches, drawings, paintings, photography, and 3D models rendered either physically or via film or Virtual Reality) and can be presented in different ways and be more or less abstract. The way visual information is presented influences one's perception—whether it's one picture, a comic strip, an animation or a video. In an animation or video, the narration is fixed in the sense that it is presented in a specific order. Meanwhile, a single picture lacks direction on how it should be read, so it is possible to start wherever we want or from whatever element we notice first (Eriksson and Göthlund, 2012; Eriksson, 2017). Comics, like an animation or video, offer a reading order and suggest a narration. However, comics include three parts that have to be analyzed by the reader: the pictures, the phrases, and the frames. The situation in which the visual information is localized also provides an entrance to understanding (Eriksson et al., 2018).

Design Thinking does not offer any methods for the analysis of final products or services. From that perspective, information design does not benefit from Design Thinking. Therefore, when it comes to the evaluation and interpretation of how an information artefact works for people, it is necessary to use additional methods and theories to determine if it is interpreted in the intended way or not. One way to determine whether or not the designed information is effective and efficient is to evaluate an individual's awareness of the information's consequences. The ability to follow instructions does not automatically mean that one understands the consequences of not following the instructions, or why it is important, in the long-term, to follow it. That is: What are the long-term consequences? This is related to *situation awareness* in the sense that the perceiver can understand what the information means both now and in the future (Endsly and Jones, 2012).

Therefore, information needs to be analyzed in relation to the situation in which it is intended to be used. The term *situation awareness* comes from military pilots, but it is relevant when it comes to the perception of information. It can be broken down into three levels: "*perception* of elements in the environment, *comprehension* of the current situation and *projection* of future status" (Endsley, 2012). In a specific and local context, individuals will probably know what to notice and, in that way, perceive relevant elements in the environment. But if the individual encounters the information in a public space, or in an unfamiliar context, the first step is to make them recognize that it is something that they should perceive.

The shape of the mind is the result of an interaction between the brain, body, and the things around us (Malafouris, 2016). How we understand something is based on "the shape of our mind", which is caused by a continual oscillation between the mental image and external input, such as things or artefact in the surroundings. Visuals in information design can be considered to be material signs, which may partly be understood as being the same as linguistic signs since they can be symbolic. However, visuals do have material qualities that make them objects in themselves. As a consequence, we can experience or interpret a visual sign as something that exists independently from that which it represents. We can also read a text, understand every word, yet not be able to grasp a comprehensible meaning from it. As mentioned before, information often persists over time, but the meaning often changes, and what remains in relation to the content depends on the situation in which people take part. While we can state that information design faces much more complex challenges than just creating comprehensible information, we can also state that information design will be more and more relevant in the future.

When it comes to information about how to assemble or use a product or service, people are typically motivated to take part in it. The challenge individuals often meet in these situations is having the ability to understand what is being communicated. There can be different explanations for this, and we often blame the instructions as not being user friendly. For instance, perhaps the information also serves other functions besides informing, e.g., legal functions (Lundin, 2019). But what does it mean to be user friendly, besides that the instructions could serve different functions? To answer this question, we need to take various perspectives into consideration. As mentioned before, this involves both socio-cultural and cognitive aspects related to physiological abilities. But what motivates a target group to take part in and use instructions should also be taken into consideration. For technical products, the instructions are often regulated by security requirements and standards. While these instructions are not always recognized by non-professional manual users, they could be crucial for professionals.

Different forms of verbal information

Like visual information, verbal information can take many different forms depending on what media is used, as well as how the message is designed and formulated. It could take the form of a recorded voice, synthetic speech, sign language, written text of different lengths, or very short text formulated in the imperative; or it could be expressed on papers, websites, walls, and floors or as Braille. Language is symbolic. Only the onomatopoetic words, i.e., words that are formed in an imitation of a sound, are iconic (e.g., *click* on the button). Sound words are dependent language. A pig's sound is spelled and pronounced "nöff, nöff" in Swedish, and "oink, oink" in English. While onomatopoetic words in sign language is iconic, which means it looks like the thing it represents. In contrast to visuals, verbal information has a reading order, and meaning in language is established through differences and through what it refers to outside itself. The symbolic character of language, then, has many possibilities in communication. With the verbal, we can communicate the abstract and that which is not currently taking place; for example, the consequences of a certain action. Its form is furthermore intertwined with its content. Rhetoric, both argumentation and style, with its tropes, forms the message and the outcome, e.g., behavioral change, of its effect on us. It wasn't always the truth itself, but the ability to convince others, that was important in the courts of ancient Greece. The style of verbal communication can then alter a meaning through its form, which Raymond Queneau's influential work, *Exercises in Style* from 1947 demonstrates through his 99 ways of telling the same story of a man seeing the same stranger twice in one day. The verbal language in information design, then, also has its challenges, since language is symbolic and can be situated, local, national, or international. Latin language in medicine is an example of the latter. Different terms for putting a nut on a screw, in the element sheets in different assembly lines, is such an example of language established inductively in the action. To be understood, information, whether it take the form of words, graphics, images or a multimodal design, must consequently be designed with an awareness of—and sometimes together with—the receivers of the information. The universal approach in verbal information design, perhaps with support from practices or research of plain or easy-to-read language, can, indeed, be combined with a situated design approach, with work-specific or langue-in-use focus. This is, however, also when it becomes hard; where standardizations, i.e., the temporarily frozen meaning, meets the living in action situated language.

Something crucial for information is that it gets attention, and for that to happen, some kind of rhetoric is often needed. However, rhetorical tropes, such as metaphors, require context to be understood. It is already understood in antiquity. However, some of the closest roots to the socio-cultural perspective on information design are considered to be communication in linguistics and semiotics. In *Mythologies* from 1957, the structuralist Roland Barthes (1969) clarifies denotative and connotative levels of cultural phenomena in his time, where the latter presupposed a cultural pattern in order to operate. The philosopher and scholar in *Historian of ideas*, Michel Foucault (1993), has also been influential; he postulates that what is written and what is thought in different discourses get their meaning due to the upholding of the order. That is why one must "be in the truth to speak the truth". From a systems perspective, the linguist Michael Halliday, who wrote *Language in society* (2009), believes that, in order to be meaningful, language must be viewed in its system, where the social structure is included. Meaning and a text's function depends on the social dimension of language. Despite all the differences between these thinkers, a common feature is that language (and, for some, also the image and other artefacts) becomes the bearer of the knowledge or meaning that the society, culture, discourse, or system maintains—and thus, also governs how we interpret information of various kinds.

Semiotic pragmatism

Following this understanding of verbal information design and the symbolic character of images, it is also important to say something further about the response of the reader (or listener). Perception is not passive. To understand our view on communication and information design, it is important to mention the relationship to reader-oriented semiotics and pragmatism. The recognition of the reader's participation, not only through perception/reception but through interpretation, is important in the communication process we underline. For a designer in a human centered tradition, it then means an acknowledgment of this activity. This follows the importance that has been given to context, culture, history, and psychology in humanities since the mid 20th century (when sociology also emerged as a science). There is a decoding and production of meaning that takes place during the reading/receiving of information. Stanley Fish, for example, speaks about "interpretive communities" (1980), and Jonathan Culler (1975) about the intersubjective semiotic systems binding language and culture together requiring a certain competence. Although this is the case with poetic literature, information and its design also have context, culture, codes, among others, for the recipient.

From the socio-cultural perspective, language and visuals are viewed as social phenomena. From this perspective, to have an effect, the designer of the information and the one who receives it must be on common ground when it comes to knowledge during the discourse. Then we come back to the discussion about the need to empathize with those who are being addressed, to which the information is being aimed at. We could even stress that the inclusion of the understandings of the information beholder is a matter of morality; it shows consideration towards the other as a condition for design in general, but especially in information design.

As we have seen, the following has been emphasized in information design in relation to Design Thinking: Information design brings the *communicative act* into Design Thinking. As a consequence, the way we understand that information design communication is not through the mechanical process school á la Shannon and Weavers' model of communication (1949) but rather, through a semiotic and pragmatic way of acknowledging both the situatedness and the generic in communication. This, in turn, brings a large portion of complexity and the need for reflection into the picture. Both information research and practice are colored by this view here. In order to evaluate the effect and impact of information design, theories from situation awareness are needed. In the next section, we will give a few examples of what this complexity and reflection can look like, addressing something generic for us all that is to be understood by specific individuals in their contexts. In addition, we will look at the ability to internalize the information and be able to use it in a relevant way, related to what is communicated; which is also a question of context and situation where individuals and groups are dependent.

An information campaign for a global pandemic

In the year 2020, the world faced a global pandemic, *Covid19*. The situation required that actions be taken in every country around the world. The coronavirus did not spread equally over cities, regions and countries. Some parts were not afflicted at all, while others were deeply afflicted. It could differ even within bigger cities, where some parts were hardly impacted. This very much had to do with socio-economic standards, education levels, and professions. On the internet, it is possible to find nearly 40,000 stock pictures, mostly pictograms, illustrating the message, "Keep distance". The design of the information posters is very similar in idea. What differs is the colors, design of pictograms, amount of verbal information, as well as the measurement system and recommended distance.

They each contain a short text requesting that people keep distance. Some of them don't tell what to do, but only show an arrow and a measurement, e.g., 6 ft. In Figure 3, from the Stockholm region, people are asked to follow the "2 meters rule", and do it "as much as possible", according to the added text. The poster also informs that this is "a way for you to save lives".

Figure 3. Poster from the region of Stockholm where people are asked to follow "2 meters rule", and do it "as much as possible".

In addition to the information about keeping distance, we find many variations of the importance of washing our hands.

The very simplified illustrations with accompanying text on posters were spread in many countries very early on in the pandemic and had a small impact on individuals to begin with. Even though the information seems very clear, it was not followed by people to the extent that one might expect. This can be explained by the big gap between the simple pictogram and the complex reality in which people recognize their needs. In the beginning, the coronavirus was abstract, and it was only when it became a pandemic that people started realizing how serious it was. However still, making sense of a pictogram, even though they are easy to interpret can be a demanding process. One has to go from a simple picture into a complex environment. Most of the pictograms show the importance of keeping distance when next to each other, and do not show how to keep distance in front of each other beyond cases of queuing. The consequences of not following the recommendations, then, had come close. The challenge was similar to that of informing about the consequences of smoking. It can't just be designed away with information. Despite information about the importance of keeping distance, the Swedish newspapers and televised news wrote and broadcasted about crowded buses, subways, and shopping centers, almost every day. How can this be explained from the viewpoint of communication and information design?

In *An Anthropology of Images: Picture, Medium, Body* (2011), the Art historian Hans Belting claims that it is the viewer who makes the picture "come to life" by "making" it in a context. When the outlook that information only becomes information when it is having an effect is introduced, it becomes a matter for information design. Like images, it is the recipient of information text who brings it to life. An advertisement or informational sign that tells us to keep our distance and wash our hands only turns into applied information when the activities are actually carried out. We keep our distance from each other and wash our hands carefully. However it is only when we have internalized the information that it will work fully; that is, when we understand the meaning and value of keeping distance from other people and washing our hands, which is also the same as embracing the consequences of not following the instructions. It is only then that the information design can make a difference. The simple pictograms that appear today in advertisements, grocery stores, and other public places about the importance of keeping distance and washing your hands are expected to be perceived and understood and thus, complied with in the same way by everyone. In fact, it is a complex relationship between the observer and the viewed, in this case the pictograms.

The complexity of using metaphors in information

Neither visual or verbal information is comprehensible outside a context or situation, whether it is a contemporary or historical context. It is a mixture of analogy and imperative when the Stockholm region goes out before Easter with these words: "Good Monday, Good Tuesday, Good Wednesday, Good Thursday, Good Friday, Good Saturday, Good Sunday, Good Monday. No ordinary Easter 2020. Continue to stay at home. Thanks" (See Figure 4).

To understand the analogy, one must know that the day was considered to be long and full of suffering and, until 1969, all entertainment in Sweden was closed on Good Friday; a Christian tradition that has its roots in the 300s and relates to the crucifixion of Christ. But for whom are all days long, and who are the ones that are expected to suffer? Those who have lost their jobs, or those who are not allowed to go to their summer houses? Those who are infected with the virus? In addition to that campaign, they also showed an egg carton with room for six eggs. Instead of a full carton, only three out of six eggs were left. This was accompanied by the text, "Keep distance during Easter". The mixing of images while aiming to strike an understanding among everyone in Sweden was complex in relation to the simple pictogram discussed above.

80   *Different Perspectives in Design Thinking*

Figure 4. One needs to be familiar with the meaning Good Friday to understand the message.

It wasn't just the region of Stockholm that was using tropes (see Figure 5).

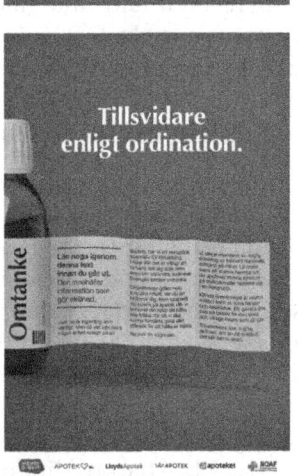

Figure 5. A medicine bottle labeled Care.

Various pharmacy companies issued a joint advertisement with the text, "Until further notice according to prescription" which alludes to prescriptions that did not yet exist at that time. A medicine bottle is visible at the edge of the picture and is labeled, "Care". The expandable user manual reads: "Read this text carefully before you go out. It contains information that makes a difference." It continues to say that there is no simple "solution", but that they have an over-the-counter alternative: "Care". It is the solution in the bottle that is prescribed to the entire population of Sweden. The solution detailed includes keeping a distance inside pharmacies, because the staff needs to stay healthy in order to be able to keep the customers well, as well as the same recommendations as those given by the Swedish Public Health Agency on social distancing, hand washing, and more. Due to the allusion, knowledge of pharmaceutical terminology is needed to understand the text. This might be clear for the majority, but the direct address, the medicine is for the reader, is clouded by the play with the term "solution", which has double meanings. In the rhetorical assonance, rhymed in the sentence that the text, "contains information that makes a difference" ("innehåller information som gör skillnad", in Swedish), one also elegantly complicates the sentence so that the "solution", that is, "care", becomes the information—which is medicine! The solution, the care, that is to be taken, and information become metaphors for one another in the ad. This style is common in advertising. And advertising for the six pharmacies is also what it is; a text where information from the Swedish Public Health Agency is used, with rhetorical subtleties, to convince us that these pharmacies still play a role in keeping us healthy when there was no medicine other than information: taking care against the coronavirus. It is an elegant, but not simple text, and the appropriation of the Public Health Agency's recommendations is, of course, good if it is followed. However it is also embedded in another for-profit case for pharmacies which, here, seems to dress in the role of the agency.

Design Thinking and human center design

In the examples from the Stockholm Region and the pharmacy companies, the recipient is required to fill in the gap between language and the image that is referred to through an understanding of rather large and complex contexts. For the information to have any effect, we must all share the same discourse and feel involved. Who is the information targeting, and how is a sense of participation created? That is, how does the information invoke the sense that this information concerns me and I have the knowledge that allows me to bring the information to life, to use Belting's terms.

If Design Thinking as a method had been used in combination with a human centered design perspective, the information designers would have done a thorough and reflective study of the various conditions for people living in different regions and the situatedness in which the information is to take place. On one hand, the information from the pharmacy companies was playful but, on the other, it was very hard to digest. Being located in the midst of information about recommendations for behavior from the authorities, illustrates yet another competence needed for the recipients of the information: the ability to distinguish who and why somebody is addressing you in the city space, in newspapers and magazines, or on your social media feed. This is also something that illustrates yet another challenge for information designers, considering they are partaking in the creation of an effect, i.e., making a difference, among many other types of messages filled with tropes striving for people's attention.

Information can be regarded as communication, which has been discussed in this chapter, though not as one-way communication. Rather, as we have stressed, it can be viewed as an interaction between the information and the recipient. During the year 2020, the design of information from the government, national, and local authorities did not change much. Months after their initial placement, the signs telling us to keep distance have been embedded in the street scene as a constant reminder. Is it the worn-down signs that contribute to what is defined as health literacy, or is it the growing number of sick and dead people? Or is the danger from the virus over, since the sign is old and has not been renewed? Once again, since information can be seen as communication, it is always a combination of several factors that makes a message come through. The worn-down information can illustrate that situatedness of information design; to be communicative is also temporal. We will come back to the temporal later.

Figure 6. A worn-down information.

The slipperiness of information

In *Shaping Information. The Rhetoric of Visual Conventions*, Kostelnick and Hassett (2003) discuss the slipperiness of conventions in graphic design. The slipperiness is also present in information as such, not only in the graphic elements that are used.

Although we share cognitive faculties as human beings, the cognitive approach cannot account for all communicative aspects of meaningful communication. The rhetorical tropes and style used, e.g., metaphors in language and images indicating "going up" or an arrow pointing up is good in certain contexts (e.g., in the stock markets), and bad in others (e.g., a sick person's temperature). Rhetoric becomes meaningful and only has an effect in a context where it is understood. The social and symbolic aspect of language that metaphors are mediating, is widely recognized today. However, these social and symbolic aspects are not static but living (including images) but the symbolic aspect, essential for human communication today, is indeed, slippery. On one hand, there is no easy answer or handbook for exactness and efficiency. On the other, there is huge creative potential in holistic information Design Thinking.

Sometimes, we need a standard to follow, and sometimes the standard is in the way. In small manual assembly teams in the manufacturing industry, where one text is produced to capture the best assembly method in this working cell at this time, the language of the operators is mixed with the company standard—otherwise, it would not be read and continuously updated by them. It is their action, their language in use.

Being a researcher and teacher in *information* design and reflecting on design thinking, and understanding language and images, as described above, both underline and alter design thinking. This is regarding the holistic and socio-cultural widening of perspectives, thinking beyond the design solution to include effects, the merging of a knowledge-information perspective integrated in information design, and the role of the designer. The role of the designer is to not only to strive for hyped innovations measured in economic success or cultural capital, but it's to aim for communication as an *enabler*, through design, of *understanding* and *effects* in the form of insight and changed behavior. This is achieved through working in design processes and placing the human way of communicating—not business or technology—at the heart of the process. Through this understanding the role of the designer becomes much less that of a romantic creative genius, and more of a knowledgeable creative enabler of change in dialogue, where all the aspects of communication and design become relevant.

Information as physical, spatial, and temporal

So far, we have mainly referenced visual and verbal information in information design. Now, we will discuss the character in which the artefacts communicate, through a comparison between the visual and the verbal in a physical environment. The environment itself influences how we perceive information. Visual information is related to spatiality, and a lot of visual information has to do with how to behave and act in space.

We live and move in different spaces. It could be spaces such as limited places in the domestic sphere, or in public spaces like workspaces. The environment is perceived through our bodies, and by moving around we see objects and people. Some objects or people are familiar to us, others are not. What we see and why often depends on the context and the situation. The discussions above regarding how and what individuals might see and understand from the pandemic campaign also need to be analyzed in relation to space. This is where the situatedness—however odd it sounds—of a holistic perspective of communication and design becomes relevant.

Where is the information displayed, and how? When walking around in a shopping center or grocery store we will become aware of the signs on the floor telling us to keep our distance. If we understand the information, and if it is crowded, we will probably come into an internal conflict if we are in Sweden, since the signs telling us to keep a distance are a recommendation, not a rule or legal command. This is yet another dimension for the information designer to consider. In countries where it is a rule, not following it is not an option, so long as we don't aim to break the rule. The distinction between information that functions as a recommendation and information that functions as a rule is not always obvious. Whether something is observed or not is often related to if it is newsworthy or not. In the beginning of the pandemic, when the posters and the marks on the floor and on the streets were new, they popped out before our eyes. By the end of the year, the marks on the street were worn down. Since the information was worn down, the embedded information is that it is from the past.

Much of the built environment contains embedded information (Bates, 2016). The purpose of a place or building, as well as its standard, could be defined as embedded information. We also use parts of or details in the environment as landmarks to orient ourselves. These landmarks could be natural or designed artefacts (Lynch, 1960). The landmark works as a tool for navigation, a kind of way showing and, in a way, it has been transformed from a natural growing tree or stone to an information artefact. Not an artefact created by a human, but something used as one.

In the book *Mind in Motion, How action shapes thought* (2019), Barbara Tversky discusses the relationship between human actions and the design of places. We organize the surroundings, interior and exterior, whole environments, cities with their different blocks, roads, parks, and so forth, based on specific purposes— living, shopping, walking, driving, for pleasure or rational purposes. Our ability to use something is how we categorize and group what we have in front of us, within our field of vision, and it's critical to our understanding of the environment. This grouping and categorization guides us in the environment, but it also limits us in relation to discovering the unexpected. The mixture of information and advertisements in public spaces is sometimes confusing, while other times it works well. Are we expected to buy something, or is the local authority just asking us to take care?

When we enter a public building, such as a hospital, library, museum or railway station, it is often for a specific reason. Even though we might think we are entering the building with an open mind (at least at the library or museum), we are often mentally occupied by some specific expectations, or busy thinking about the purpose of our visit or the departure time of the train. Once again, we go back to empathizing with the individuals for whom we are designing a specific piece of information. Let's look at an example of a train station that is being reconstructed. How can we inform travelers departing and arriving at a train station undergoing extensive reconstruction where to go and how to find the right track? Who is the target group or users? The answer you will often get is, "everyone", and that, therefore, it is impossible to have a specific group in mind. We argue that this is not the case, since there are several ways to group users. In this case, we are dealing with a specific context and *situation*. This refers back to the discussion on information in relation to the pandemic. If the information is to be effective and perceived as meaningful by people, where it is physically located, how it is spatially placed in relation to the place and, finally, whether the information is temporary or permanent, is crucial. It could be temporary but present for a relatively long-time span, in which case the information design would need to be updated in order to become visible again to the people who regularly use the space where the information is posted. Otherwise, the effect of the information runs the risk of losing its communicative ability, since people will ignore it.

Conclusion

The semantics of information does not sustain over time, since meaning is something that occurs, or is created, among the recipients of the information. The aim of information design is to design information so that it is communicated to the recipient in a meaningful way. We have argued that communication is a two-way act in which individuals and groups need to take an active part in order to turn the information into a meaningful message. To make this possible, both the designer and the recipient need to reflect on the intention of the communication. The challenge in information design is that individuals and groups perceive and understand information differently, depending on the individuals' or group's motivation, experience, and previous knowledge. People's motivations could be dependent on aspects such as socio-economic and cultural background, gender, age, ethnicity, and education level.

However, communication has several different dimensions. It also deals with rhetorical tropes, semiotical signs, and cognitive aspects. Every single dimension contains several spectra in which communication is the aim for each part. Information design must be discussed in relation to physical, spatial, as well as temporal aspects. We need to understand the information design product or services and, from the product, how to behave or act. Therefore, the relationship between the bodily experience and the sign is important. One has to internalize the information in order to create a situation awareness where the sign can transform into something meaningful that makes sense and affects an action or behavior.

There is ongoing discussion about the value of Design Thinking, with focus on its very simplified understanding of a design process and, especially, as a business model. Therefore, Design Thinking needs to be discussed from a human centered perspective that does not simplify the process but takes a holistic perspective of what is going to be designed and how it will be used and by whom. At the same time, the role of information design research and information design needs to shift from delivering comprehensible information to being a part of product design and product development and contributing to workplace design that facilitates communication and the use of instructions during the performance of work tasks. Looking at the communicative aspect of design at large, not just product language but how products interact with us and vice versa. In fact, the relationship between the observer and the viewed is complex. Making information come to life is a multi-layered process in which individuals and groups need to notice, perceive, analyze, de-code and be able to use the information in a meaningful way. That process is what leads to insight and/or action.

References

Bærenholt, J.O, Bücher, M. and Damm Scheuer, J. 2014. Perspectives on design research. *In*: Simonsen, J., Baerenholdt, J.O., Büscher, M. and Damm Scheuer, J. [eds.]. Design Research. Synergies from Interdisciplinary Perspectives. Routledge, Abingdon, Oxon.

Barthes, R. 1972. Mythologies. Harper Collins, Canada Ltd., Canada.

Bates, M.J. 2016. Fundamental forms of information. Journal of American Society for Information and Technology, 57(8): 1033–1045.

Belting, H. 2011. An Anthropology of Images: Picture, Medium, body. Princeton University Press, New Jersey, USA.

Brown, T. 2009. Change by Design. Hoe Design Thinking Transform Organizations and Inspires Innovation. Harper Collins Publisher, New York.

Chimuanya, L. and Aijboye, E. 2016. Socio-semiotics of humor in ebola awareness discourse on Facebook. pp. 252–273. *In*: Taiwo, R., Odebumni, A. and Adetunji, A. [eds.]. Analyzing Language and Humor in Online Communication. IGI Global, Pennsylvania, USA.

Culler, J. 1975. Literary competence. pp. 131–152. *In*: Culler, J. [ed.]. Structuralist Poetics; Structuralism, Linguistics and the Study of Literature. Cornell University Press, Ithaca.

Di Salvo, B., Yip, J., Bonsignore, E. and Di Salvo, C. 2017. Participatory Design for Learning. Perspective from Practice to Research. Routledge, London/New York.

Endsly, M.R. and Jones, D.G. 2012. Designing for Situated Awareness. CRC. Press, Boca Raton, Florida, USA.

Eriksson, Y., Johansson, P. and Björndal, P. 2011. Showing Actions in Pictures. Information Visualization. London UK. 13–15 July.

Eriksson, Y. and Göthlund, A. 2012. Möten med bilder. Att tolka visuella uttryck. (Encounter pictures. To analyses visuals). Studentlitteratur, Lund, Sweden.

Eriksson, Y. 2016. Technologically Mature but with Limited Capabilties, 2 nd International Conference of IT for the Aged Population, HCI 2016 International, Springer Verlag.

Eriksson, Y. 2017. Bildens tysta budskap. Interaktion mellan bild och text (The silent messages of visuals. The interaction between pictures and text). Studentlitteratur. Lund, Sweden.
Eriksson, P.E., Swenberg, T., Zhao, X. and Eriksson, Y. 2018. How Gaze time on Screen Impacts the Efficacy of Visual Instructions. Helyion 4.
Fish, S. 1980. Is there a text in the class? The Authority of Interpretive Communities. Harvard University Press, Cambridge, Mass. and London, England.
Foucault, M. 1993. Diskursens ordning (L´ordre du discourse, Installation lecture, Collége de France, 1970). Brutus Östlings Bokförlag, Stockholm.
Gunn, W. and Donovan, J. 2012. Design Anthropology. An Introduction. Design and Anthropology. Ed. Wendy Gunn and Jared Donovan. Routledge, London and New York.
Halliday, M.A.K. 2009 Language in Society. Continuum, London.
Hård af Segerstad, P. 2009. Kommunikation och information: en bok om människans förmåga att tala, tänka och förstå (Communication and information: a book about human being capability to talk, think and understand). Liber, Stockholm.
Kostelnick, Ch. and Hassett, M. 2003. Shaping Information. The Rhetoric of Visual Conventions. Southern Illinois University Press, Carbondale.
Lundin, J. 2019. Shaping thought through action: A Study of the use and Design of Technical Information. Doc. Diss. Mälardalen University, Eskilstuna/Västerås, Sweden.
Lynch, K. 1960. The Image of the City. MIT Press, Mass. Cambridge, USA.
Malafouris, L. 2016. How Things Shape the Mind. A Theory of Material Engagement. MIT Press, Mass. Cambridge, USA.
Martin B. and Hanington, B. 2012. Universal Methods for Design. Rockport Publishers, Beverly, MA, USA.
Queneau, R. 2012. Exercises in Style. A New Directions Book, New York.
Simonsen, J. et al. (eds.). 2014. Situated Design Methods. Cambridge, Mass. and London England: The MIT Press.
Tversky, B. 2019. Mind in Motion, How Action Shapes Thought. Basic Books. New York.

CHAPTER-6

Gendered Design Thinking
So-called Logic of People and Things

Åsa Wikberg Nilsson[1,*] *and Yvonne Eriksson*[2]

Introduction

In the broadest sense, there are almost no limits to what design can be. Thinking of design as all the human-made graphics, things, interactions, and systems that people interact with every day, makes it a powerful arena and practice. We often seem to consider design as something that is passive when we humans interact with or otherwise experience it. However, as anthropologist Marshal Sahlin (2013) states, designed things are never semantically sterile or ideologically passive; they are always— intentionally or not—inscribed with the meanings and values of the culture that produced them, and the culture in which they are situated and used. In this text, the overall objective is to draw attention to the lack of understanding that exists when it comes to the powers, privileges, norms, and logic that uphold structural inequality in design. Here, we conceptualize such intentional and unintentional doings as *gendered design thinking*, meaning that, without a deeper understanding of how the unequal cultures and collaborations are formed, designers risk upholding inequality.

In design, there has been great interest in some of the renowned design thinkers' ideas of design thinking; for example, Nigel Cross' (1982) portrayal of "designerly" ways of knowing, Bryan Lawson's (1997) ideas on how designers think (Lawson, 1997), and, not the least, the design consultant Tim Brown's (2008) accounts of design thinking in itself. Rarely, however, do such discussions involve gendered ways of thinking, i.e., the ideologies, norms, and assumptions about people and things that

---

[1] Industrial design, Department of Social Sciences, Technology and Arts, Luleå University of Technology.
[2] School of Innovation, Design and Engineering, Division of Information Design, Eskilstuna/Västerås Mälardalen University.
* Corresponding author: asa.wikberg-nilsson@ltu.se

operate in and around design. For example, the aesthetics, feminist, and emotional theorist Carolyn Korsmeyer (2004) describes how traditional concepts such as genius, beauty, and aesthetic perception are qualities that are closely linked to gender. This involves, for example, paying attention to the role gender plays in the realization and use of design(s), as well as who is privileged to be a genius or creative person, the power of defining design values, and the different gendered factors that constitute the design culture itself. Author and journalist Katrine Marçal (2020) states that gendered thinking is the reason for missed innovation opportunities. She exemplifies this using everything from electric cars and suitcases, to space suits and AI (artificial intelligence). Through these examples, she illustrates how norms and values of masculinity and femininity have been characterizing what investments are worth considering , or otherwise .

Norms and values can change, and it is possible to reinvent the future for those who dare to think beyond prevailing gender norms. A point of departure for such strivings is to turn to social theory, to understand the concepts of gender, power, privileges and other human sense-makings. In this chapter, we introduce some of the basics to provide a foundational understanding and shared language for the constitutions of gendered design thinking: the so-called logic of people and things.

"Doing gender"

The concept of gender was first introduced in the 1970s to differentiate the biological sex from the socially and culturally created roles and identities that are respectively associated with women and men. However, it is not easy to distinguish the biological from the cultural and, perhaps, this is especially the case in design. This means that gendered design thinking extends far beyond discussions of female designers or design for women. Instead, it involves a deeper understanding of the fundamental gendered norms, power structures, logic, and meanings that fluctuate in different historical and cultural design contexts. Generally speaking, a gender perspective comprises the analysis of differences in power and privileges for women and men in all societal areas and arenas. Gender studies are often motivated by the endeavor to create democratic, human-centered societies. However, despite the fact that gender equality and reduced inequalities have been part of our global goals for decades, they still haven't been realized.

One reason for this could be that gender is embedded in the power structures and so-called logic of the systems in which we humans operate: in the ways we identify ourselves, in our everyday interactions, in power relations, and in society as a whole. Ethnomethodology scholars Candace West and Don Zimmerman introduced such ideas through the concept of "doing gender" in 1987. This concept involves the theory of gender inequality as something we humans create and reaffirm in everyday human actions and interactions. According to this view, there is a need to problematize the prevailing idea of female and male as naturally defined binary categories, represented as each other's opposites, resulting in different mental and behavioral inclinations. Such created opposites—called *dichotomies*—are rooted in the biological sexes, which makes the so-called logic about them both basic and persistent. However, rather than being a fundamental quality of women and men, "doing gender" involves the theory of a constructed gendered system of thought. This system, and its logic, can be easily recognized in contemporary media. Women are often represented in passive, social roles involving relationships, friendship, and family, while men are represented in important work roles, both in blue and white-collar work, and supplementary activities (Rose et al., 2012). Such stereotypical gender roles are also manifested in certain video games, where female characters are represented as highly sexualized beings, while male characters are represented as strong, hyper masculine and aggressive beings who lack emotions. From this perspective, the visual communication of stereotypical gender roles and identities contributes to reaffirming people's "gender logic".

The historian Yvonne Hirdman discusses the gender system as being based around the principles of segregation and hierarchy (Hirdman, 2001). *Segregation* means that the binary categories of woman/female and man/male are divided and created as each other's opposites. This creation results in women and men being associated with different arenas and with doing different things at work and home, which involve a certain "logic" of natural predispositions. Gender theories involve the problematization of this being about biological predispositions and, rather, it is viewed as a social and cultural act of valuing things differently for women and for men (West and Zimmerman, 1987; Hirdman, 2001). An example of gender segregation is the design business itself, which is predominantly white and male-dominated (Design Council, 2005; Allen, 2013; Reimer, 2016). One consequence of this is the associations between craft, skill, and masculinity, which are reinforced in design practice (Reimer, 2016). There are also historical remarks of gender segregation in design areas, where female designers were told that they were best at the areas that weren't so esteemed and well-paid (Buckley, 1986). This involves gendered design thinking that results in the categorization and prioritization of certain types of design, design areas, distinct styles and movements, as well as different modes of production which are inherently biased against women and serve to exclude them from design and its history. This does not mean that women haven't been designers or that some part of their biological constitution isn't suited to designing. Rather, as design scholars Karin Ehrnberger, Minna Räsänen, and Sara Ilstedt (2012) conclude, the consequence is that women's influence, ideas, and designs have been thoroughly depressed.

The other gender system factor is *hierarchy*, and it relates to power. According to Hirdman, this entails that what is viewed as a male arena, and that which is related to being a man, means more, is valued higher, is viewed as natural, logical, or simply the norm. On the other hand, what is related to being a woman or female is viewed as the secondary, or the deviant, which must be clearly separated from the norm. This can, for example, be observed in certain products made for women where the attribute "for women" is communicated on the label, while similar products made for men do not have any such attributes highlighted; they are for the man, i.e., using the word "man" is analogous to using the word "human", or those who are the norm. The sociologist Karl Marx (1818–1883) noted that people's logic is fundamentally wrapped up with their power position; hence, the privileged in any society control what ideas are accepted as "logical". According to the sociologist Allan Johnson (2001), power, privilege, and oppression only exist in social systems through and by the participants who are a part of these very systems. This relates to gender aspects of womens' and mens' abilities to influence, but also to other aspects such as race, class, functional-ability, language, age, and sexuality. Power and privilege are hence linked to the social, political, and cultural logic in which gendered thinking contributes to strengthening existing privileges and power; consequently, maintaining an advantage for certain social groups. Sociologist Patricia Hill Collins also argues that gender, class, and race are highly influential in creating such ways of thinking or "logic". She problematizes our ways of creating knowledge and points at the impossibility of separating yourself from what is being studied (Collins, 2000). In summary, we have to change our ways of gendered thinking, as related to people and things, in order to change the ways people and things are considered. Instead, we must build our ways of knowing upon lived experiences. Collins argues that it is only those people who know from personal experience which should be part of "the knowers". She also argues for the use of dialogue and personal pronouns in the stories being told, thus emphasizing that all knowledge is value-laden and should be the subject of empathy, as emotions indicate a genuine belief in the argument. Since knowledge, in this view, builds upon personal experiences, all actors should be able to express their beliefs, values, and ethics. Collins questions which ways of thinking could actually contribute to the goals of equality: the ones that deny ethics and moral accountability or the ones that request it?

Science and technology scholars Ann-Jorunn Berg and Merete Lie (1995) state that, in a society based on power, hierarchy, and inequality, gender is not negotiated on equal terms. In this sense, one danger of overlooking gendered design thinking is that, instead of representing empathy and human-centricity, the design process and its results reflect and uphold patriarchal power structures and privileges. A result of this is that, when faced with designs that don't suit their needs or interests, women, as well as other disregarded users, tend to avoid them, and thus male designers can continue to disregard them (Berg and Lie, 1995). Although Berg and Lie first made these observations in 1995, they still hold true in society today. Gendered ways of thinking are created and recreated both on individual and structural levels, and involve a continuous reconstruction of what is considered to be "logic" about women and men. Therefore, gendered design thinking sometimes contributes to insidious leaps of mind, and results in designs that represents biases rather than human experiences.

To understand how and why design incorporates bias and barriers against certain user groups, there needs to be a problematization of users as imagined, rather than truly empathized, by the designer. Science and technology scholar Madeleine Akrich (1992), for example, describes the inscription of representations of users and use upon artefacts, in the form of scripts. These scripts attribute and delegate specific competences, actions, and responsibilities to users and the design itself. If the user representation fails to match the actual user, the design will probably fail. The user representation is, of course, based on the designer's "logic" of how people are. If the designer doesn't interact with and identify specific needs that either or both women and men might have, for example, small hands that make it difficult to grasp large tools, then the resulting designs will not fulfill such needs.

Sex vs. gender

The relationship between the concepts of sex and gender is complex, and most simply explained as the biological versus the social and cultural. A basic definition is that biological sex is what is determined by chromosomes, hormones, and external and internal genitalia. However, research indicates that the biological sexes are not that easy to distinguish from each other; a person could have a mix of, e.g., female genitalia and male chromosomes, for example, hermaphroditism, transgender bodies, manly elderly women, and menstruating and breastfeeding men (Money et al., 1955). Already then, in 1955, the boundaries between female and male bodies were perceived as uncertain and dependent on external factors such as the level of civilization, division of labour, climate, habits, lifestyle, upbringing and, not in the least, morality. Theories about gender as something that is changed and shaped by external circumstances enabled, among other things, the comparison of different groups of genders and diversity based on class, age, and race (Larsson, 2000). Still, however, we tend to relate to the initial discussions about biological sex from back in the 1970s, rather than to the social and cultural "doing" of gender.

To complicate things—and despite gender being a social construction that varies with culture and tradition—it often relates to the biological sex, such as expectations about how to behave and dress according to if you are a woman or a man, and what to work with or be interested in, whether you are a woman or a man. Children are fostered to meet these expectations from an early age, and children's toys often emphasize biological sex in relation to play and profession (Ehrnberger et al., 2012). Illustrations in children's books often contribute to this by showing boys in scenarios where they are playing with airplanes, cars, and pretend to be firemen, while girls in the illustrations (which are less frequent) are depicted together with horses, dogs, cats or flowers (Eriksson, 2013). This goes hand in hand with the patriarchal tradition of regarding women and children as a part of nature, and men as a part of culture (Buckley, 1986). In this sense, these stereotypes impact the physical spaces that women and men occupy, whether at home or at work, as well as their occupations, and their relationships with design.

Of course, it is your individual right to decide not to accept or adapt to gender norms. However, this is not up to each individual to define, rather, it is defined by the people we interact with. Gender is not created in a vacuum, it is created in interaction with the surrounding people and things. People can either follow the norms or violate them. However, the violation of norms could, in some subgroups lead to the creation of new norms. Moreover, from there, the new norms will be adapted to by society. One example of this is that women today can wear pants and suits without violating norms, which was definitely not the case less than one hundred years ago. This illustrates that norms are changeable.

One aspect that matters in some contexts is biological sex and age, especially when it comes to medical treatment. Some biological functions and bodily aspects differ between men and women, something that has been identified in healthcare and medical research. Even if we take gender perspectives into account in design, meaning we support change in behavior and are open for possibilities of cross-gender behavior, it is important to recognize biological differences and needs. Studies for example indicate that the muscle mass of men is larger than that of women, and the average man is longer and bigger than the average woman (Tilley, 2001). These differences are not true for all men and women, and of course we can find many examples showing the opposite: all men are not big and strong, some are shorter and have less developed muscle mass, and so on. However, if we ignore these aspects, designers will also ignore that different sizes and weights of, tools could hinder some women and small men to perform activities they want to do. The importance here is to recognize that for example, an electric drill, does not always need to be heavy, but could still be effective. Less heavy or big tools should neither automatically have less performance, nor be painted in pink.

The challenge for designers is that, on the one hand, they need to recognize the biological differences between women and men, and also consider women's constitutions, values, and experiences—something that hasn't historically been the case. And, on the other hand, they need to avoid designing based on stereotypical beliefs of women and men. To clarify, not all women like pink and wear high heels, that is a somewhat simplified stereotypical expectation, as is the idea that all men have large hands and huge muscles so that they can use heavy cumbersome tools. Finding this balance between identifying both women and men's values and experiences, and avoiding outcomes that contribute to stereotypical behavior or value of the same, is difficult—but necessary—if an equal society and a truly human-centered design practice is to be realized. Then, and only then, we might be able to impede the gendered and often stereotypical separation of women and men's predispositions, and instead talk to people and design for human beings with all their different individual experiences and aptitudes. To counteract the doing of gender means to be in dialogue with history, culture, politics, and to continuously strive for visions of an equal society. Human-centric design thinking means that design is made for different humans and their needs, and also their different capacities. To be a human means a lot, and is not a uniform concept; we are all individuals who come in different ages, heights, weights, and with different physical or cognitive limitations.

Technological developments enable people to do things they could not do previously. Harbor work is no longer heavy; almost all lifts are handled by cranes steered by humans. Before the power steering was developed, only strong people (mainly men) were able to drive tracks, but today it is an easy task to steer a track—no muscle power is required. These are only a few examples of where design, together with technology, has made a huge difference. If gender is acknowledged, not only as size differences between men and women, but also as social and cultured values and power, design can be an effective and efficient tool for bridging the gap between people and enhancing their ability to perform outside the gender norms.

Gendered design history

It is not possible to talk about *a* design history. There are many different histories, depending on the area of design and what part of the world the history covers. In this section, we will provide a brief history of the gender differentiation that emerged in the western world in early modern times in relation to industrialization. During industrialization in the mid 19th century, the cities in Europe started to grow, as did the middle class which began consuming the mass-produced products. One of the business ideas of the time was to produce gender and age diverse products—everything from small everyday artefacts to furniture. The products produced included things such as special hairbrushes for men, and for women, pocket mirrors for women to keep in their handbags, and wristwatches for women and pocket watches for men. Special porcelain for children was also produced, and decorated with suitable motifs from the world of fairytales. Even furniture was produced that, for the first time, catered to children, such as small chairs and tables. For grown-ups, desks were specially made for women and for men (Forty, 1986). These early initiatives set the agenda for gendered design thinking which, more or less, has lasted up until today and, to some extent, is still being reproduced.

Gendered design thinking is not only about differentiating between the predispositions of women and men. It was, and still remains, a question of what is considered to be good and bad taste or style, and who is privileged to be part of that discussion or decision. Related to this, the design critic Philippa Goodall (1990) describes design as a discourse: a set of practices and knowledge that constitutes the whole field of design in a given society. Since the practice of design and designers is implicated in the processes of creation, the production and exchange of everyday objects, designers' notions of women and femininity—as well as men and masculinity—are, in this sense, both influenced and formed by design.

The design historian Penny Sparke describes the shift in the relationship between consumers and producers in relation to power. In the Victorian days (1830–1900), taste was entirely assigned to women; for example, in their domestic activities of decorating the house, and so on. There was also a dichotomy between good and bad taste, in which ornaments and decorations were both feminized and devalued as bad taste. In the public domain, however, it was the opposite. The power to value taste in the public domain was not considered to be taste but, instead, it was seen as men's rational logic of what works. In this design thinking discourse, function was valued over form, less was considered more, and industrial mass production and high-tech materials were highly valued. The right to value a design object was, hence, transferred from the consumer, who was often a woman, to the controlling hands of the professional (male) designer, who formed and defined a design language based on masculine terms, with the consequence that other values were degraded and soon diminished (Sparke, 2010).

An example of this is the shopping cart, an everyday object which, today, most people would not associate as being either female or male. However, this was not the case when it was introduced in the USA in the 1930s. Author and journalist Katrine Marçal (2020) explains that the idea originated from the observation that people could only carry a certain amount of groceries and, if they could transport more things they would, of course, consume more too. Hence, the shopping cart was introduced, allowing people to keep their hands free and shop for a lot of things at once. Yet, men refused to use the new product. They viewed it as a personal insult, an object that insinuated they weren't strong enough. In effect, before the shopping cart could become the everyday object it is today, the gender norm that indicated that the act of rolling your goods is "unmanly" had to be confronted. Actors (men) were hence hired to stroll around the stores with shopping carts, and, with time, the gender norm was changed.

The masculine taste has been assigned higher value in different design contexts, and so-called gender-neutral artefacts are often designed to suit men. Perhaps this isn't always explicit, but the outcome is still the same. Women have adapted to the male sphere and, today, are accustomed to using artefacts that were previously intended for men. But men are not adaptable in the same way. This can be explained by the power hierarchy between men and women. To step up and perform like a man is not regarded as a loss of prestige, but for a man to behave in a way that is only expected of women entails a serious loss of prestige. This is also mirrored in the design of everyday artefacts.

Visual communication and gender

Visual communication signifies everything that is visible; that is, everything that can be perceived and communicated visually, such as gestures, mimics, and movement. Visual communication is part of the visual culture which includes visual artefacts, but also the artificial and material environment surrounding humans. The visual culture consists of everyday objects, cars, various kinds of tools, as well as art, and fashion. Objects and the built environment (and, of course, nature itself) communicate different values that will affect individuals and the interaction between individuals and groups. Often, the values of a designed object or environment are embedded and not explicitly expressed—or people don't always recognize it—but still, they influence the understanding or use of the object or environment. Visual communication belongs to materiality, except for colors which are simply visual. The mind and things interact in a continuous process, not as isolated independent entities (Malafouris, 2013). What we take part of in the visual culture will impact how we think and act, and vice versa. Our mental images affect how we perceive and interpret the world.

Much of visual communication is mediated by visual representations. A representation is a re-presentation of a thought, object, or environment. How the re-presentation is designed or what section of an image is used in photos depends on both the aim of the picture and the tradition. This decision could be made deliberately or unintentionally, but it will affect the communication. Communication needs to be understood as an interaction between individuals situated in context and temporality (see Eriksson and Carlsson in this book). It is not possible to fix the meaning of a visual representation since different individuals will perceive and interpret it from their horizons of experience and understanding. Since seeing is an active process, it is something we do, and continuously learn to do:

"Seeing the world is not about how we see, but about what we make of what we see" (Mirzoeff, 2015, p. 73).

Media, culture, and communication scholars Marita Sturken and Lisa Cartwright describe these differences as *seeing* and *looking*, where seeing means to notice, while looking is an activity (Sturken and Cartwright, 2009). Looking is an activity that not only involves an interpretation of what has been seen, but also a reflection of what one has looked at. A reflection could refer to both something that is mirrored on a surface, and a serious thought or consideration. If we reflect upon what we have seen or looked at in the mirroring sense, it often means that we are imitating its meaning without considering what we have seen. We imitate things such as gestures and movements without thinking about how they fulfill gender expectations. Much of that which is gendered in visual communication is the result of mirroring rather than serious consideration. Designers and consumers are influenced by each other and we often disregard how gender is communicated in everyday products and packaging. Colors, symbols, materials, and shapes are parts that we associate with gender, not necessarily consciously, but unconsciously. Product packaging is a way to communicate with expected consumers and also pick out the presumptive buyers. The packaging not only displays the product, but it often shows a picture of the expected user and how it will be used. In addition, most products today are also present on the internet. It's not only toys that are given a gender stamp, but this also takes place with products for adults. However, there is a tendency to play down gender advertisements and product information that were previously gender coded; for example, for vacuum cleaners, blenders, and lawn movers. Hardly any users are visible in advertising online, but when the lawn movers, for example, are displayed in a context they are more often than not used by men.

Visual communication in design thinking consists of several layers. It begins with figuring out how to visually formulate sketches in the early phases of a design process: *How will the formulation affect the ability to empathize with the expected user in a human-centric design process? And what is the understanding and interpretation of the shape and colors, if they are included?* In a design process, the gendering of a product starts from the moment the very first line in a sketch is formulated in interaction with a mental image of the coming product. The choice of materials and colors usually comes later in the process, and is affected by the shape of the object. These aspects support the gender differentiation in visual culture and visual communication. We see gender, we think it, and we (re-)create it. To uncover the traditional gender norms, we need to deconstruct our ways of looking and thinking, and learn how to look and think in new ways.

Forms and norms

The design scholar Marcus Jahnke (2006) states that design reveals our norms and values: the dichotomy between binary gender roles such as male and female, where the former is valued higher, is reconstructed in the form, color, and material of an object. The binary gender roles, for example, describe different characteristics. If you look in a dictionary, you might see that the definitions of the term *male* include qualities such as "vigor" or "boldness", and the term *female* is associated with attributes such as "small in size" and "tenderness".

The philosopher and gender theorist Judith Butler (2011) refers to gender as something that is performative and, as such, is both stable and unstable. According to this view, an act that is being done iteratively, over and over again, is performative if it produces a series of effects. Gender identity is hence constructed through iterative gendered acts, and constituted in our everyday human lives and interactions. Gender norms, what we consider to be female or male, are unstable and changeable. Gender norms can be analyzed in representations of symbols, materials, forms, colors, and action elements prescribed for a female or male market. In many everyday products, designers have thought about the customer's gender. This is sometimes subtle and goes almost unnoticed as we humans seem to have normalized or grown accustomed to such gendered design thinking. In some design areas, this is an aspect that is being used as a playful design element, as "gender benders". An example of this is the perfume produced by fashion houses such as Jean Paul Gaultier. The fashion design rebellion in the 1980s and 1990s consciously played with stereotypes. Jean Paul Gaultier used the shape of a male torso for men's perfumes with the names *Ultra Male* and *Le Mâle*. Meanwhile, the bottles of their women's perfumes took the shape of the female torso decorated with different details and colors, and were given the names *Amazon, Le Belle, Cabare* and *Classique*. However, the consumer of the perfume does not necessarily see it as a play with gender roles, or as homoerotic ideals.

Gender is also performed in artefacts that are not always thought of as gendered, such as robotics. However, forms often follow norms, especially norms that are related to gender. In the field of engineering design, there is a tendency to ignore gender since it is often stated that it is irrelevant. Thus, robots' visual designs influence our perceptions of them, and we may even apply gender stereotypes to them, as people tend to think that robots with "male" and "female" torsos are respectively suitable for traditionally male and female tasks (Bernotat et al., 2017). In addition, research indicates that people respond to single communication modalities (face, head, body, voice, locomotion), but this is not necessarily related to a higher degree of anthropomorphism (Tsiourti et al., 2017). These response patterns will probably influence the acceptance of robots in various contexts, and an intriguing question is whether robots will cement traditional stereotypical gender roles and codes, or break them. The relationship between bodily representations and gendered expectations of robots has been apparent, at least since Rotwang made a female robot that met traditional ideals of what it meant to be a beautiful woman in the movie *Metropolis* (Eriksson, 2018).

The design and culture scholar Magdalena Peterson-McIntyre (2013, 2018) states that designs cannot be seen as innocent objects, but rather, designs are active participants in the reproduction of gendered thinking through their form, materials, colors, and associated commercials. The things themselves contribute to normalizing certain kinds of seemingly desirable femininity or masculinity. They affect us, and thus become co-creators of gendered design thinking. In this sense, design representations conceal power by shifting gender to a matter of consumer choice, rather than making it a matter of the designer's intention and accountability. But is it really what the customers want or is it, rather, the designers' gendered "logic" about the user?

If you take an everyday product such as razors, and critically analyze how it contributes to the reproduction of gendered norms—for example Gillette's design of razors targeting a female or male market—it is quite apparent in its gendered design thinking (Wikberg Nilsson and Jahnke, 2018). Consequently, the razor for men is made of black and metal-colored plastic, giving the illusion of being steel, and has a sharp, straight form with supplementary line decorations to illustrate speed. Its package fulfills the impression, featuring greenish lines associated with lasers that seem to vibrate out of the razor, relating it to technology. It clearly takes an active stance. The razor for women is pink and has a curved shape that can be associated with the female body. It has a somewhat stable form, made from a mat plastic material. There are no extra features added to this razor; it takes a rather simple and non-technical stance. Compared to the men's razor, the added feature for women is the packaging text—something that is seemingly not necessary for the other one. Even the product's names contribute to a re-construction of gender identities. The razor for men is called Mach^3 Innovation, representing power, technology, and action, and the one for women is called Venus, the very symbol of beauty, love, and fertility. Correspondingly, the marketing of these products contributes to a reproduction of gendered design thinking. The razors for women are advertised using a photo of a nameless woman in the bath, tenderly smiling and looking away from the camera. In contrast, the razors targeting the male market are promoted by a well-known athlete with his arms crossed, staring directly into the camera, illustrating notions of actionability. A contemporary analysis carried out at the beginning of 2021 looking at razors for men and razors for women illustrates the same pattern.

Interaction norms

It is difficult to even imagine a world without design, without the signs, objects, interactions, or systems that constantly surround us humans. We need them in order to live and, willingly or not, we are surrounded by their culturally and socially inscribed meanings. Related to this is design scholar Donald A. Norman's (2002) notion of user experience, which emphasizes that the purpose of a "thing", whether that thing is visual, physical, a service, or a system, is to serve the user. In this sense, a high-quality user experience goes far beyond giving users what they say they want. For example, empathy mapping is a design tool that is said to support designers in developing an understanding of users (Gibbons, 2018). It involves visualizations of what designers know about users based on interviews, observations, and user testing. However, here's the deal with empathy: even if you make the metaphorical effort of putting yourself in someone else's shoes, it will still be your own feet you are walking around with. Also, there is a distinction between what it takes "to be with" versus what it takes "to be like" someone; i.e., there's a shift from transferred to shared and lived experiences (Bennet and Rosner, 2019). Neither design thinking or user experience approaches seem to recognize gender and intersectional elements as important for better addressing user insights. It might, for example, be difficult for some users to express their needs. An example may be that a man may find it difficult to express that he would not be able to handle a certain technology, since the gender norm prescribes men as tech-savvy. Likewise, a transgender person may have difficulty defining the experience of not belonging to the normative binary gender roles and, hence, would neither subscribe to the tech-savvy hero, or a pink-it-and-shrink-it strategy. A black woman might not even be asked about her experiences with soap dispensers, as some contemporary sensor technologies don't distinguish dark skin (Shareable, 2015). If she had been acknowledged as an important source of user insights, such inattentiveness might have been avoided.

There is this thing called *user insight drifts*. Design consultant Tim Brown's (2005) statement about designers needing to translate user insights is one example of this. However, computer scientists and information system designers Mike Robinson and Liam Bannon (1991) question representations, as designs pass between many different professionals like marketers, managers, economists, tech-specialists, designers, among others, and each of the groups have their own "logic" and specialized language that can be referred to as a semiotic community. When designs pass between such semantic communities, user insights might get lost in translation, resulting in a disparity between design expectations and the final outcome. Robinson and Bannon make a distinction between description and interpretation, hence, denotation and connotation in semiotic term (Barthes, 1994), and emphasize that, if this is not considered, the final design outcome may come as a surprise for everybody involved and result in dissatisfaction.

A general design strategy for developing products, services and technologies is to *design for everybody*, without having any particular user group in mind. However, to think of the user as "everybody" has proven to be an inadequate strategy by science and technology scholars Nelly Oudshoorn, Els Rommes, and Marcelle Stienstra (2004). In this sense, the designer configuration of the user as everybody leads to solutions that are biased towards young, white, and well-educated male users, which also reflects the power structures within design and among designers. Women are often in the minority in design and tech businesses. Studies have demonstrated strong alignments between technology and masculinity (Cockburn and Ormrod, 1993; Wajcman, 1991, 2007; Berg and Lie, 1995; Faulkner, 2000), associating technology with what it means to be a man (Connell, 1987). As a result, some user insights might be overlooked, and contribute to male biases in technology products:

> "Tech companies are often dominated by male engineers or at least by their way of thinking" (Schroeder et al., 2012).

The previously described science and technology scholar Madeleine Akrich's concept of gender scripts outlines how designers' user representations also shape designs, as they inscribe the user interaction with certain actions and behaviors. Thus, the designs, in turn, shape the user context, inscribe their interactions, and serve as ubiquitous reminders of gendered design thinking, and who is included or not. Users can, of course, choose to refuse, that is, not subscribe to the proposed visuals, products, services, or systems. Scripts, however, both invite and constrain; they are mediators between humans and their world. In this sense, design plays an important role in shaping individual identities, social structures, and cultural values, and vice-versa. Social norms contribute to shaping and creating designs based on ideas, and values of people, things, and behavior. Designs are embodied by their users. As humans create relationships with things, they define the user experience by directing the interaction, and play a background role in contextualizing the experience and shaping meanings in the human relationship with the environment.

If we are interested in human-centricity in design, a consequence is that "we cannot be satisfied with the designer's or user's point of view alone. Instead, we have to go back and forth continually between the designer and the user, between the designer's projected user and the real users, between the world inscribed in the object and the world described by its displacement" (Akrich, 1992). The design scholar Donald Schön (1983) describes seemingly similar situations as the designer's reflective dialogue with the design situation. This is, of course, one way of counteracting the prevailing understanding of the user and situation at hand. However, it is also plausible that such dialogical exchange needs to be immersed and situated into "a particular", as feminist theorist Donna Haraway (1988) describes that there is no neutral or distanced position before or after. It is instead, a process that starts with challenging one's own preconceptions and biases, i.e., our horizon of understandings, our logic; and through continuously encountering others, new ways of thinking about people and things that may emerge.

To understand how and why gendered design thinking contributes in setting up barriers for certain user groups, there is a need to reconsider the mutual interrelation of gender and design. Designs cannot be viewed as separate from humans and society, they are not merely technological or social. Instead, there is a need to address the combinations of denoted and connoted meanings and representations of symbols, objects, actions and systems of thoughts (MacKenzie and Wajcman, 1999). The current digitalization is one example of this. Artificial intelligence (AI), information technology (IT), and other technologies make some human tasks easier and faster, but things that happen fast and without serious consideration might go wrong. As the machine learning scholar James Zou and science historian Londa Schiebinger (2018) state, there is a need to improve the fairness of data and AI; should the data be representative of the world as it is, or of a world that many would aspire to? One example is Amazon's attempt at designing an AI based recruitment tool to ease their recruitment processes and job application processing. In machine learning, there is a need for large data sets that AI applications can be trained upon. Amazon receives a lot of job applications and used this as training data. However, the problem was that the majority of the applications came from men and, also, the outcome of the recruitments were usually men, illustrating the gendered thinking which results in tech companies alienating women in recruitments (Wynn and Correl, 2018). Nevertheless, this meant that the AI recruiting tool learned—on its own—that men were preferable, and screened out all the women who applied for jobs in any tech-related area, no matter how experienced they were (Dastin, 2018). In Amazon's favor, the project was terminated immediately.

Design and gender systems

Design scholar Richard Buchanan (1992) states that the complex systems and environments for living, working, playing, and learning need to express a balanced, functioning whole. Yet, scholars Tora Friberg and Agneta Larsson (1999) describe how there are also gender-coded areas and domains in urban design; for example, the domestic/familial sphere is coded as a women's domain, while the public sphere is coded as gender-neutral but, in reality, plays out as a men's domain. In this sense, public places are opportunity spaces, comprising expressions and representations of both physical and social norms in society. Traditional city halls, in some cases, materialize gender norms through dark oak materials, large and heavy front doors which make you feel small when entering, and a lot of gold frames with pictures or paintings of the men in power, carrying both explicit and implicit codes of who is allowed. So-called modern building materials such as concrete, steel, and glass are characterized by the industrial revolution ideals of technology and rational production standards, leading to precision and uniformity in forms and materials that are easy to mass produce (Alves Silva et al., 2016). As previously described, typical gendered thinking of women and men is reproduced in forms and materials. Products intended for men express speed, technology, and strength, with straight and angular shapes. Products intended for women express sensuality and playfulness. There are, however, not that many examples of the latter in urban places and spaces. In Friberg and Larsson's view, it is therefore necessary to explore the *materiality norms*; what coded messages do the forms and materials communicate, and who is inscribed in the messages? In this sense, exploring places and spaces based on the *mental maps* of different people is also important. This can reveal user experiences of places where people of different genders, ethnicities, ages, physical abilities, and so forth, don't feel included, feel insecure—or feel the opposite, safe and included—and offer a better understanding of what contributes to such experiences. There are also a lot of oblivious places and spaces, so-called *white spots*, which affect people's everyday lives but are strangely overlooked.

One example of such inattention is winter road maintenance. In the city of Karlskoga in Sweden, the local municipality used to remove snow based on the hierarchy of first, attending to the main roads, second, attending to the roads around large work places, and then taking care of walkways and other places (Björkman, 2018). However, when the municipality managers took part in an equality leadership program, they started to question such gendered thinking: they realized that women walk, ride their bikes, or use public transportation more often compared to men. They also experienced that it is far more difficult to walk in 10 centimeters of snow than to drive through it. Also, there are three times more pedestrians than drivers that are part of accidents who, for example, slip and require hospital care. Most of these pedestrians are women. Health care and sick leaves due to slip and fall accidents cost four times more than the city's winter road maintenance. The local municipality managers analyzed this from a gender perspective and realized that they had prioritized men; their first priority was the main roads, used by men in cars, which lead to larger male-dominated workplaces. They didn't perform such gendered logic consciously, they just did what they had always done without asking questions. However, through analysis they understood that they had to re-think the whole system. Hence, the first priority for the snow maintainers became schools and preschools where parents drop off their kids, and then large workplaces—but now also female-dominated ones, like hospitals and other municipality facilities. Only after this are large roads and other areas maintained. In their experience, it's not more expensive to re-think previous priorities (and hence privileges), but it is more inclusive in terms of who has access to the resources; it benefits everyone—children, elderly, functionally-disabled, and others—regardless of gender (Björkman, 2018). The lesson here is that how our cities are designed affects how people can live their lives. Genuine, human-centric design thinking is therefore to listen, empathize, and include a diversity of people's needs, experiences, and insights of a particular place or situation.

Conclusion

Design and design thinking continue to expand meanings and representations in society. What Buchanan (1992) stated some thirty years ago still seems relevant today: "There is no area of contemporary life where design—the plan, project, or working hypothesis which constitutes the "intention" in intentional operations—is not a significant factor in shaping human experience" (p.8). To paraphrase Buchanan, this also necessitates a need to challenge gendered design thinking, the so-called logic that contributes to our notions and values of both people and things.

Thus, there is a need to challenge the *visual communication* that represents the stereotypical gendered thinking and design culture of the past which contributes to shaping people into different, predefined binary gender roles. There is also a need to attend to the representations of physical, psychological, and social relationships between humans and *products* based on an awareness of norms, stereotypes and biases of people of different genders, races, sexualities, classes, ages, abilities, and so forth. This necessitates further inquiries into how products are formed, and how they form us. *Interactions* and services need to be challenged in terms of how human-centered design thinking can contribute to making such user experiences more intelligent, meaningful, and satisfying for a diversity of people; for example, through the active participation of people with lived experiences. The complex *systems and environments* for living, working, playing, and learning need to be challenged in terms of who participates in framing the central idea, thoughts, or values, so that they actually contribute to being balanced and diverse, functioning wholes. This concerns the role of design in sustaining, developing, and integrating human beings into broader ecological, social, and cultural environments, shaping the designs when desirable, and carefully nurturing the designers' logic of people and things to let a thousand flowers bloom. Visual, physical, and interactive things, services, systems and environments are different dimensions of design thinking. However, through reconsidering and rethinking of all these dimensions in terms of gender, privileges, power, and oppression, and as part of forming a balanced and functioning whole, they can also be the very ground for genuine, human-centric inventions. In doing so, the field of design can take the next step in challenging inequality and power imbalances in all societal sectors, and truly aligning design thinking with the intention of being human-centric.

References

Akrich, M. 1992. The description of technical objects. pp. 205-224. *In*: Bijker, W. and Law's J. [eds.]. Shaping Technology - Building Society: Studies in Sociotechnical Change.
Allen, K. 2013. What do you need to make it as a woman in this industry? pp. 232–253. *In*: Ashton, D. and Noonan, C. [eds.]. Cultural Work and Higher Education. Basingstoke: Palgrave MacMillan.
Alves Silva, M., Ehrnberger, K., Jahnke, M. and Wikberg Nilsson, Å. 2016. NOVA—Tools and Methods for Norm-creative Innovation. Stockholm: Vinnova.
Barthes, R. 1994. The Semiotic Challenge. Berkley, CA.: University of California Press.
Bennet, C.L. and Rosner, D.K. 2019. The Promise of Empathy: Design, Disability, and Knowing the "Other". CHI 2019. Glasgow, Scotland, UK, pp. 1–13.
Berg, A.-J. and Lie, M. 1995. Feminism and constructivism: Do Artifacts Have Gender? Science, Technology, & Human Values, pp. 332–351.
Bernotat, J., Eyssel, F. and Sachse, J. 2017. Shape It—The Influence of Robot Body Shape on Gender Perception in Robots.
Björkman, L. 2018. Uppföljning av jämställd snöröjning. Karlskoga: Karlskoga kommun.
Brown, T. 2008. Design thinking. Harvard Business Review June: 84–92.
Buchanan, R. 1992. Wicked problems in design thinking. Design Issues, 7(2): 5–21.
Buckley, Cheryl. 1986. Made in patriarchy: Toward a feminist analysis of women and desgin. Design Issues, 3(2): 3–14.
Butler, J. 2011. Gender Trouble: Feminisim and the Subversion of Identity. Routledge.
Cockburn, C. and Ormrod, S. 1993. Gender and Technology in the Making. London: Sage Publications Ltd.
Collins, P.H. 2000. Black Feminist Thought: Knowledge, Consiousness, and the Politics of Empowerment. New York: Routledge.
Connell, R.W. 1987. Gender & Power. Cambridge, UK.: Polity.
Cross, N. 1982. Designerly ways of knowing. Design Issues, 3(4): 221–227.
Dastin, J. 2018. Amazon scraps secret AI recruiting tool that showed bias against women. October 10. https://www.reuters.com/article/us-amazon-com-jobs-automation-insight/amazon-scraps-secret-ai-recruiting-tool-that-showed-bias-against-women-idUSKCN1MK08G.
Design Council. 2005. The Business of Design. London: Design Council.

Ehrnberger, K., Räsänen, M. and Ilstedt, S. 2012. Visualising gender norms in design: Meet the mega hurricane mixer and the drill dolphia. International Journal of Design, 6(3): 85–98.
Eriksson, Y. 2013. Teknik och genus i bilder som vänder sig till barn. Bilden av ingenjören, Ed. Yvonne Eriksson and Ildiko Asztalos Morell. Stockhom: Carlsson Bokförlag.
Eriksson, Y. 2018. The perception of aging and use of Robpta. Springer International Publishing AG, part of Springer Nature 2018 J. Zhou and G. Salvendy (Eds.): ITAP 2018, LNCS 10926, pp. 30–39.
Faulkner, W. 2003. Teknikfrågan i feminismen. In Vem tillhör tekniken? Kunskap och kön i teknikens värld. Lund: Arkiv.
Forty, A. and Cameron, I. 1986. Objects of desire: design and society since 1750. London: Thames and Hudson.
Friberg, T. and Larsson, A. 1999. Om manligt och kvinnligt i planeringens könsneutrala värld. Nordisk Arkitekturforskning, 1999(2): 33–44.
Gibbons, S. 2018. Empathy Mapping: the first step in design thinking. January 14. Accessed 09 2020. https://www.nngroup.com/articles/empathy-mapping/.
Haraway, D. 1988. Situated knowledge: the science question in feminism and the privilege of partial perspective. Feminist Studies, 14(3): 575–599.
Hirdman, Y. 2001. Gender: about the changeable forms of the stable (In Swedish Genus: om det stabilas föränderliga former. Malmö: Liber.
Jahnke, M. (ed). 2006. Formgivning/Normgivning. Göteborg: Center för konsumtionsvetenskap, Göteborgs universitet.
Johnson, A.G. 2001. Privilege, Power, and Difference. New York: McGraw-Hill.
Korsmeyer, C. 2004. Gender and Aesthetics: An Introduction. London: Routledge.
Larsson, M. 2000. Ett ängsligt sysslande med könens ordning: Medicinska tolkningar av kroppen vid 1800-talets mitt, i Bedrägliga begrepp: Kön och genus i humanistisk forskning, Opuscula Historica Upsaliensia 24, Ed. Gudrun Andersson, pp. 51–72.
Lawson, B. 1997. How Designers Think: The Design Process Demystified. Oxford: Architectural Press.
MacKenzie, D. and Wajcman, J. 1999. The social shaping of technology. Buckingham, UK: Open University Press.
Malafouris, L. 2013. How Things Shape the Mind. A Theory of Material Engagement. Mass. Cam.: MIT Press.
Marçal, K. 2020. Att uppfinna världen. Hur historiens största feltänk satte käppar i hjulet. Stockholm: Mondial.
Marx, K. and Engels, F. 1978. The german ideology. pp. 146–200. In: Tucker's, R. (ed). The Marx-Engels Reader.
Mirzoeff, N. 2015. How to See the World. Pelican Publishing Company.
Money, J., Hampson, J.G. and Hampson, J.L. 1955. An examination of some basic sexual concepts: the evidence of human hermaphroditism. Bull John Hopkins Hosp., 4: 301–319.
Norman, D. 2002. The Design of Everyday Things. New York: Basic Books.
Oudshoorn, N., Rommes, E. and Stienstra, M. 2004. Configuring the user as everybody: gender and design cultures in information and communication technologies. Science, Technology & Human Values, 29(1): 30–63.
Peterson McIntyre, M. 2013. Perfume packaging, seducation and gender. Culture unbound. Journal of Current Cultural Research, 5: 291–311.
Peterson McIntyre, M. 2018. Gender by design: Performativity and consumer packaging. Design and Culture, the Journal of Design Studies Forum, 10(3): 337–358.
Reimer, S. 2016. Its just a very male industry: gender and work in UK design agencies. Gender, Place & Culture, 23(7): 1033–1046.
Rose, J., Mackey-Kallis, S., Shyles, L., Barry, K., Biagini, D. et al. 2012. Face it: the impact of gender on social media images. Communication Quarterly, 60(5): 588–607.
Sahlins, M. 2013. Culture and practical reasons. Chicago: University of Chicago Press.

Schön, D. 1983. The Reflective Practitioner. New York: Basic Books.
Schroeder, K., Mosegaard Vilhelmsen, S., Sogaard Jensen, C., Jacobsen, M., Norager, R. et al. 2012. Female Interaction Strategy. Aarhus: Design-people.
Shareable. 2015. Racist Whites Only Soap Dispenser. November 3. https://www.youtube.com/watch?v=kycsE2djcc4.
Sparke, P., Massey, A., Keeble, T. and Martin, B. 2009. Designing the Modern Interior: From the Victorinas to today. Oxford, UK.: Berg.
Sturken, Marita and Lisa Cartwright. 2009. Practices of Looking: An Introduction to Visual Culture. New York: Oxford University Press.
Tilley, A.R. 2001. The measure of man and woman: human factors in design. Hobroken, NJ.: John Wiley & Sons.
Tsiourti, C.H., Weiss, A., Wac, K. and Vince, M. 2017. Designing emotionally expressive robots: a comparative study on the perception of communication modalities. pp. 213–222. *In*: Wrede, B. and Nagai, Y. (eds.). Proceedings of 5th International Conference on Human Agent Interaction HAI'17. New York: ACM Press/Addison-Wesley, 2017, Bieldfeld, Germany.
Wajcman, J. 1991. Feminism Confronts Technology. Cambridge: Polity Press.
Wajcman, J. 2007. From women and technology to gendered technoscience. Information, Communication & Society, 10(3): 287–298.
West, C. and Zimmerman, D.H. 1987. Doing gender. Gender and Society, 1(2): 125–151.
Wikberg Nilsson, Å. and Jahnke, M. 2018. Tactics for norm-creative innovation. She-Ji: The Journal of Design, Economics, and Innovation, 4(4): 375–391.
Wynn, A.T. and Correl, S.J. 2018. Puncturing the pipeline: do technology companies alienate women in recruitment sessions? Social Studies of Science, 48(1): 149–164.
Zou, J. and Schiebinger, L. 2018. Design AI so that it's fair. Nature, 559: 324–326.

CHAPTER-7
On Design Dialogues
Their Roots, Features, and Usage

Ulrika Florin

Introduction

In certain design and development processes, there are good reasons to explore user insights in a more extensive way, especially if the design affects people's everyday lives, work-lives, or both. User insights are particularly valuable when it comes to designs that impact society at large.

It could be the design of a public place, a hospital or a workplace, or design that's less tangible in nature, such as infrastructure, communication systems, and working processes. In any case, when artefacts, processes and systems impact many people, it means, quite certainly, that the problem behind the designed solutions is quite complex. If the solution, then, is poorly grounded, it can also lead to unforeseen implications on a systemic level. For such designs, the presentation of a set of different proposals to which potential users and stakeholders are invited to give their opinions, is not enough. Instead, there is a need to open up for dialogue and engagement so that the concerned users and stakeholders are given the opportunity to contribute their knowledge and experience to the design process.

Let's say, for example, that an extra entrance to a subway station needs to be built due to the rising number of people using it. The problem here, is figuring out how the new entrance can be fit into the existing environment and how, and if, it will be accepted by the public, since it has to be built in a beloved green recreational area of the city. All other location options have already been ruled out, with consideration for how the subway track layout interacts with the placement of sewers, electric power lines, and buildings. Over and above this, there are also other

School of Innovation, Design and Engineering, Division of Information Design, Eskilstuna/ Västerås, Mälardalen University, Sweden, Email: ulrika.florin@mdh.se uncertainties in this scenario which must be taken into consideration; some with systemic implications that can be related to diverse sustainability aspects. Firstly, the building of a new entrance in this location might impact the city's air quality, as several trees must be felled to make room for it. This would reduce the city's green lung, unless it's compensated for by planting new trees elsewhere in the city. However, green areas in densely populated areas not only serve as lungs, as in clean air to breathe, but they also serve as an important counterbalance to the stressful pace of the city and are, therefore, indispensable to human health (Hartig et al., 2020). Secondly, we have the crowding scenario to consider, which calls forth and increases the risk of chaos. However, crowding in itself also includes other systemic aspects such as contagion exposure, which is a problem planners and designers now pay more attention to—awoken by the Covid19 pandemic. This unforeseen event has placed the prevention of disease transmission high up on both local and international agendas and, more precisely, crowding as a specific threat in this respect. Such scenarios must be thoroughly prepared for. Therefore, design methods that can oscillate at holistic systems-oriented angles on problems are needed in order to picture and handle the prerequisites beyond shape, size, construction, costs, aesthetics, among others during planning. Scenarios, like the pandemic, are time related, ambiguous in nature, and primarily come into sight during events—and can easily be overlooked during planning. Thus, temporal and unknown aspects must also be weighed and balanced during design and planning processes to prevent mistakes from being made.

Examples involving systems related risks were voiced already in the late 1980s by the design thinker Victor J. Papanek (Papanek, 1988). In his article, "The Future Isn't What It Used to Be", he begins with a critique that is directed towards both technologically driven design approaches (with systematic and predictable design processes and their attempts to rationalize and construct), and intuitively driven design approaches (with methods and processes that address sensations and emotions). However divergent, both design processes were, as Papanek reasoned, "neglecting" knowledge derived from other domains. Papanek's article demonstrates that enormous amounts of data regarding humans and their environments was available at the time, but unused in (many) design processes. Useful knowledge from areas such as ergonomics, ecology, archaeology, psychiatry, cultural history, anthropology, biology, ethology, and cultural geography, were ignored in the planning (Papanek, 1988). A specific example from Papanek's text is "The microbes in the towers", which tells a story about what neglecting this kind of information can mean. It is about two office and residential towers that are placed close to each other and built out of dense materials with unopenable windows. The buildings with integrated heating and cooling systems reuse the same air over and over again—which quite quickly resulted in the air carrying fungi, bacteria, and particles which were circulated around in the buildings. Moreover, the fresh air intake of one building was placed close to the other building's ventilation exhaust. This meant that the fungi, bacteria, and particles from one building could enter the other and spread there as well (Ibid, 1988). Reading this story, it's easy to picture how the air travels via the buildings' technological heating/cooling systems into people's heart-lung systems, in and out, back and forth, over and over again.

Papanek's critique in the 1980s stressed that designers need to incorporate knowledge on environmental and human factors in their planning and designs. His talks and works have informed design practitioners and policy makers, and increased their awareness of the consequences of not bringing other knowledge into the design processes. The design practitioners and policy makers have since adopted more transdisciplinary approaches that help them to tackle design problems in a more informed way. Today, however, design methods and philosophies have advanced even further and new methods have emerged, and are being developed, that enable a complex problem to be tackled as a complete whole, which also includes the experience and knowledge of stakeholders and users.

Design Dialogues—a means for complex design

One way to stage and carry out design processes that can tackle challenging and complex problems is to stage so-called *design dialogues*. This interactive design method was developed to aid the exploration of problems with high complexity within the built environment (Fröst, 2004; Fröst et al., 2017). Moreover, the method is also used as a means of investigating and promoting change and development processes within organizations (Florin and Söderlund, 2018). Design dialogues are one way to understand complex problems together with users and stakeholders, and are typically arranged as a series of workshops, for which specific artefacts are designed and used as tools in exploratory group dialogues. The usual practice is to build from one workshop in order to inform the next and, through such iterations, make knowledge about a problem denser and maturer. Design dialogues can also be staged as single events with parallel sessions that tackle different angles of a single problem; but with a design that supplements the exploration of the whole system within which the design problem is identified. The *for-purpose-designed* artefacts are what differentiate the design dialogues from other interactive group dialogue methods. The design artefacts are literally in center, to stimulate interaction in more ways than solely via verbal language. It is a process designed to aid the discovery of user experiences and perceptions, and to put them into play, and on display. The method offers a democratic and inclusive form of exploration, where users and stakeholders interact with each other, the designers (researchers), and the artefacts. This method has the potential to inform design and development processes in multimodal ways.

The design dialogue and its roots

Design dialogues, as a method and strategy, have been developed by practitioners and researchers from the fields of architecture, art, and design. The method is strongly rooted in the material culture that acknowledges skills or practical knowledge related to exploring, inventing, creating, and doing. This core places design dialogues in the *design thinking* field via Nigel Cross's notion of "designerly" ways of knowing, thinking, and acting (Cross, 2001, 2007). Nigel Cross's extensive design research has, among other things, articulated the nature of design activities and expertise and developed design thinking by highlighting design cognition as a particular aspect of our human intelligence. However, this is clearly related to what the philosopher John Dewey advocated for in 1929: a new relationship between science, art, and practice, and knowledge which is achieved through a kind of art directed toward change (Dewey, 1929/1958). Moreover, this development must also be attributed to Donald Schön, a philosopher and specialist in urban planning, who highlighted the professional's ways of learning and "reflecting-in-action" (implicit, direct, and embodied), as well as their explicit and deliberate "reflecting-on-action" (Schön, 1983, 1987).

Ontologically, the design dialogues belong to the family of design-driven dialogues which, in turn, can be placed in the landscape of *participatory design* methods, where *co-design* and other well-known approaches are included (Sanders and Stappers, 2008). Participatory design methods are set forth by scholars as a set of exploration methods where designers and researchers come together with users and stakeholders to jointly develop a design object or thing, often by means of lo-fi prototyping. Participatory (user-involving) approaches, as such, are methodologically related to practice-based research, focusing on exploring, elucidating, and benefitting from practical experience. Michael Polanyi voiced the idea of tacit (implicit) knowing in the 1960s, stating: "Our body is the ultimate instrument of all our external knowledge, whether intellectual or practical. In all our waking moments we rely on our awareness of contacts of our body with things outside for attending to these things" (Polanyi, 1966). In other words, that practical experience cannot be disconnected from our bodies—no experience can, for that matter.

Scholars have also noted that participatory design is influenced by (interactive) *action research* in the sense that the methods not only serve as a means of studying a situation, but also serve as an attempt to change it (Bannon and Ehn, 2012). This dimension is also true of the design dialogues, with the two folded aim of understanding users' needs and learning from their experiences, as well as involving them in the actual development or change process to some degree. *Human centered design* and *inclusive design*, as a stance, can be viewed as integrated perspectives in the design dialogues; meaning that users are thought of as the center for development, in contrast to more technology driven design processes.

Developed in Scandinavian traditions, participatory design started in the 1970s and is grounded in the simple, but intricate, idea that the people affected by a design should also have a say about it (Björgvinsson et al., 2012). Similar thoughts also motivated the development of design dialogues, but exactly when this method—or more precisely, the naming of it—occurred is difficult to distinguish. In any case, design dialogues seem to have developed in relation to the larger domain of interactive research approaches. Nevertheless, it is clear that the methods have grown in parallel, with impulses from different traditions and praxis', and not exclusively from the area of design thinking. The development is also inspired by systems thinking—specifically, the soft systems methodology—which was initiated in the 1980s by the management scientist Peter Checkland, with a focus on "practical use in real-world problems" (Checkland, 2000). Later on, in light of the need to explore stakeholder values and boundaries in order to challenge marginalization, the systems thinker Gerald Midgely argued for a more pluralistic use of methods from the systems literature and beyond—in an effort to create a flexible and responsive systemic action research practice (Midgely, 2000, 2008). It is clear that the development of interactive methods stretches over several research areas and relates to the challenges of contemporary society, with the need for sustainable solutions to increasingly complex world problems. This was the nudge that created the shift towards participatory and systems connected exploration methods.

However, explanations of this type of method development, with participatory influences, can also be traced all the way back to the 1930s and a sort of pre-start of the Bauhaus movement. During this time, the boundaries between different creative practices started to dissolve. From there, the method evolved further in relation to Nordic Functionalism, with its relation to social democracy and the Swedish welfare ideology, *folkhemmet*. To simplify, the development first arose with the dissolution of design practitioners' inner boundaries and the rise of the habit of working in teams. Then, in the next step, development continued through different attempts to integrate users in design and change processes. Both of these adaptations were driven by the endeavor to understand and solve design problems in a more informed way. Informed in the sense of learning through and including user insights and perspectives during the design process, as well as learning from each other. However, learning from and taking care of people's experiences put a demand on the designers (and researchers) to develop additional new skills; including a sensibility for the scaffolding of dialogues that can invite different expressions, and the ability to highlight and incorporate user insights in the design process.

These kinds of projects are often driven in cross-functional and/or multi-actor teams. A cross-functional team consists of employees from different functional areas in an organization mixed together in a project working group; even if each member simultaneously belongs to their functional team, i.e., production, maintenance, sales, *and others*. In multi-actor teams, different partners in a project, instead, bring complementary types of knowledge—such as scientific, practical, or other types of experiences—in order to make a joint effort in project activities from beginning to end.[1] This is to ensure that the solutions are more ready to be put into practice, and can meet actual needs. Its purpose is also to develop well-grounded and sustainable solutions that supplement more technical and aesthetic design aspects in a project.

The first project described as "real participatory design" emerged in Scandinavia in the 1970s, and was prompted by the introduction of computers in workplaces (Bannon and Ehn, 2012). In turn, this shift towards more computerized ways of working led to discussions about the value of labor and what kinds of skills and knowledge society rewards. This was also expressed with renewed interest in the theoretical conception of practice and personal involvement as a prerequisite for knowledge. As a consequence, the debate on practical knowledge (know-how, implicit knowledge, knowledge-in-action, or tacit knowledge), once again, became a subject of intense debate among scholars in the 1980s and 1990s (Göransson, 1983; Molander, 1992, 1996). The fear articulated was whether or not it would be possible to substitute professionals with artificial intelligence, an issue that still arises from time to time. Anyway, this called for a renewed focus on understanding knowledge, and Polanyi's idea of tacit knowledge was once again brought to the table; pinpointing the human body as the core instrument for connecting with external knowledge—whether intellectually or practically.

The characteristics of design dialogues

As stated earlier, design dialogues are preferably conducted as a series of workshops. It is central that both users and stakeholders meet in interactive (multimodal) dialogues for-purpose-designed artefacts. What characterizes the design dialogues in relation to other related methods,

---

[1] Many of the Horizon 2020s calls for funding required to apply a so-called multi-actor approach (MAA). Pointing towards interactive and demand-driven innovation and design (Horizon 2020).

more specifically, is the distinct use of practical design methods and techniques. Designerly ways of knowing and doing are fused to the exploration method, but are also integrated in the forward processing of possible solutions. In this respect, the method is closely related to the work of architects, designers, and artists, in which the knowledge and skills related to the making of (design) objects and processes reside at the core. When specific workshop material is to be produced, for instance, practical skills such as how to cut cardboard or form plaster into models, as well as skills in sketching and drawing techniques—both analogue and digital modes—are of vital importance. Another characteristic that design dialogues focus on is the creation of something concrete. That is, the design or alteration of a specified space, building, district, system, or process—something that can be outlined and decided upon. And, as Peter Fröst and his colleagues have clarified, design dialogues "are not only about agreeing on what qualities you want to achieve, but also about how these qualities can actually be expressed in a specific solution" (Fröst et al., 2017).[2] The process is built around a series of time-concentrated and content-structured activities, workshops, with the purpose of developing visions and goals through joint exploration and design activities. These workshops are partly designed with a commitment to creating a common understanding of a design problem and, partly, to actually creating the design proposal together (Fröst, 2004).

By working in this fashion, a design team will become acquainted with the problem and, while scrutinizing it and developing a possible solution, develop an understanding of its full complexity, involving different knowledge, experiences, and horizons. Solutions will, thereby, also be anchored in involved organizations and society, since users and stakeholders have already been involved in the process from an early phase.

Tangible material

The for-purpose-designed artefacts—tangible material—used in design dialogue workshops may consist of visual scenarios, photos, maps, layouts, models, as well as interactive games, with explicitly designed game boards, tokens, and reference figures. Consequently, the preconditions for conducting the method are that at least one person on the team, who is planning and scaffolding the workshops, needs to have the required multimodal skills. This means that they not only need to be able to produce the artefacts, but they must also have the ability to interpret the altered scenarios, i.e., analyze results from one workshop in order to fruitfully design material for the next. Additionally, they also

---

[2] Quote translated from Swedish to English by Ulrika Florin, 2020.

need to have the ability to present the outcomes in meaningful ways for the intended receivers. The resonant idea behind the design dialogue is to be able collect and bring forth both concrete and implicit information via the altered workshop materials (altered designs and scenarios), as well as from what is reflected verbally in the dialogues. So, it is not only about the doing aspects, but it is also about the ability to read and relate to objects from abstract domains, such as blueprints, maps, and scale models. This includes the understanding of relationships, positions, functions, and different scales, both in digital and analogue modes; as well as the ability to recognize, interpret, and sort out measurable and subjective relationships. This also includes keeping the significance of parts and wholenesses in the investigated problem within sight. The latter is directly related to a soft system understanding.

Artefacts in group processes

The value of using artefacts in group processes has been investigated from different perspectives over the years. In the late 1980s, the sociologist Susan L. Star and the philosopher James R. Griesemer introduced the *boundary object* theory, a pioneering concept which emphasizes that different kinds of artefacts can serve as a crossing point (or interface) between different social worlds and practices (Star and Greisemer, 1989). According to this theory for heterogeneous collaboration, the artefacts are considered to be resonant entities. In recent decades, the knowledge-bridging characteristics of the artefacts themselves have been more closely studied with regard to what aspects can support or hinder communication, with emphasis on type, style, scale, referential aspects, relationships, and user domains, including facilitation aspects (Eriksson and Florin, 2011; Florin et al., 2012; Florin et al., 2013; Florin, 2015). The importance of verbal instructions and gestures as supplementary guidance for understanding pictures and diagrams is also emphasized (Kang and Tversky, 2016; Eriksson, 2017). Also, studies on group design processes, regarding how a group understanding of key project aspects occur, and how the under-development-artefacts (and similar products already on the market) stimulate a shared overview and foster a group understanding (Olsson and Florin, 2011). Moreover, research highlights *visual awareness* as a specific knowledge requirement when exploiting designed artefacts in a communication aiding capacity in cross-functional teams (Florin and Eriksson, 2020).

Complex problems

Now, recall the design problem that was presented in the introduction of this chapter, about planning a new entrance to a subway station. The complexity of this design problem calls for a more extensive exploration, and the need to capture unforeseen angles of the problem by involving users and stakeholders. There are additional motives for involving users in such projects, namely, the need to anchor the new design so that the public can accept it. Accordingly, the motivation for the involvement of users and stakeholders has multiple layers. In this scenario, the first layer is the possibility of thoroughly investigating various aspects of a coming design together with the actual users and stakeholders, to include their perspectives and knowledge already in the early development phase. The second layer can be articulated as more of a means of establishing a coming change. Both are equally needed in the endeavor to explore and solve complex problems for public environments.

In literature on both design thinking and systems thinking, the exploration and solving of complex, "wicked" problems is highlighted as the reason for their significance (Buchanan, 1992). The motive behind both these research (and practice) areas is the pursuit to find the best possible solutions to complex problems—to the extent of what we can learn about a specific problem at a certain time. A central property of complexity is that we can only know what is possible to investigate in a certain moment. This means that taking the conscious stance, that things will possibly change, is central to the design of investigations aimed at tackling complex problems. Scenarios might be altered due to a change of political governance, or it could be nature itself that impacts scenarios with, for instance, higher water levels than expected. Another example is the Covid19 pandemic, during which almost all long-term development projects were altered, and borders between countries temporary closed and opened again, affecting changes in predicted logistics and travel patterns. As a result of the pandemic, we also altered our ways of working and how we meet, and people were forced to use digital devices more extensively. So, even if forecasting methods are used to predict future scenarios, there can be systemic uncertainties that are uncounted for. In any case, both design thinking and systems thinking are about how to understand and explore a wholeness rather than its parts. Hence, to date, there is only a limited amount of research highlighting the two areas as integrated approaches, which explores principles (Jones, 2014) and discusses the experience of field-use with this integrated perspective (Pourdehnad et al., 2011; Florin and Söderlund, 2018; Buchanan, 2019; Söderlund et al., 2020). The development of the design field and aspects of design history, with the numerous connotations the term *design* has had over the years, is one explanation. These connotations, or meanings, reflect different intentions connected to designed artefacts; including their aesthetic properties, forms, and functions, and how these aspects have been altered as a consequence of new production methods which allow for larger product series. In parallel, the more artisanal, smaller product series (and single objects) have developed into something that carries an implicit exclusivity, sometimes also signed piece by piece by the designer. In the last four or five decades, additional understandings of design have been added which relate to the activity of designing services and systems. However, already in 1992, Richard Buchanan illustrated this development by summing up design in four areas related to designers' work and works: "the design of symbolic and visual communication", "the design of material objects", "the design of activities and organized services", and "the design of complex systems or environments for living working, playing, and learning" (Buchanan, 1992).

Nowadays, the meaning of design that relates to the act of designing services and systems is highlighted as an advantage for companies or organizations when they develop new segments in their businesses. This particular meaning of design has also been spread by specific interest organizations, such as the Swedish Industrial Design Foundation (SVID) in the first decade of 2000, and is spreading in parallel within research and education in innovation, design, and management. At the same time, the more common understanding of design as something limited to form, function, and aesthetics, is still rather dominant among people at large. This means that it can sometimes be hard to explain how design methods, such as design dialogues, can be helpful in complex design processes, depending on the organization's awareness of this contemporary meaning of design.

As in the case with design, there is no unified understanding of what constitutes systems, even if the systems thinking literature seems to have a more unified connotation of the term *system*. Systems thinking deals with wholenesses and their parts, and the interconnectedness between parts and wholenesses. In any case, central to systems thinking (or doing) is tracing and defining boundaries related to complex problems that need to be tackled (Midgley, 1992, 2000). Hence, the boundaries, rigidly outlined, can also exclude aspects that might lead to a limited understanding of a problem. Therefore, Midgley (1992) advocates for capturing the insights emanating from marginal "elements" (people and things), which are neither fully included nor excluded from the system. Midgley states:

> [—] we cast light on a system by defining a boundary, the 'sharp line' between the region of light and what lies in the dark is, in a sense, an artifact: we also have the possibility of looking for gray areas in which marginal elements lie [—].
>
> Midgley (1992)

Midgely's idea is to develop a language for primary and secondary boundaries, but also for what lies in-between. The primary boundary, the most obvious, might be placed around a defined organization, an ecosystem, a society, or a planet *et cetera*. The secondary boundary allows the recognition of associated elements outside the defined system, that are seen to affect it, the relationship between things and people outside the system, which still affect it. "Elements seen to be lying between the two boundaries are marginal to the system" (Ibid., 1992).

Systems, as such, can be physical, mechanical, philosophical, social, organic or hybrid. They can be open, closed, dynamic, peripheral, subordinate or central. They can be an assemblage of things united by some sort of regular communication, symbiosis, or interconnectedness. They can be natural and/or human made, physical and/or conceptual, closed and/or open, and include static and/or dynamic components (Tien and Berg, 2003). Systems can be differently detailed with regard to incorporated elements—composed of people, processes and products—that also can intersect natural and artificial domains. In terms of characteristics, systems can be formed as processes, modeled by components that transfer specific features which relate and interact with the system as a wholeness. Systems, accordingly, are filled with details and formed with or without human involvement. They can consist of both processes and products or things. Characteristics, relationships, and interactions with other systems influence how a system (or systems) can be perceived, described, and altered. The same might be translated into design: a design can be filled with details, formed with or without human involvement; a design can be both processes and products; and characteristics, relationships, and interactions with other designs having an impact on how a design (or designs) can be perceived, described, and altered (Florin and Söderlund, 2018).

Examples from practice

The most documented use of design dialogues can be found within architecture and planning, with examples from the development of hospitals and other care environments, and school and office buildings (Fröst et al., 2017; Fröst, 2020). In this area, the adoption of design dialogues came relatively early, which can be explained by the fact that care environments, as such, are complex. Architectural projects in this domain are, therefore, characterized by the need to include various health professionals in the creation of solid and sustainable concepts, so that these types of buildings can serve both the requirements of patients and caregivers. In recent years, caregivers' working methods have also been altered towards more patient-centered care philosophies. This means a change in requirements which, preferably, are identified together with users (Ibid., 2017).

Urban planning is another area where the use of design dialogues is slowly being established—sometimes performed as artist driven projects that involve habitants in the transformation of residential areas—

especially in projects that aim to transform the marginalized outskirts of cities, including the refurbishment of so-called million program areas.[3]

Adjacent fields, such as workplace design and research on work environments, are also beginning to apply design dialogues to explore more fluctuating needs from an office worker's perspective. The use of design dialogues in this field emphasizes the employee's involvement when developing workplaces and spaces, by putting attention on the work's concrete, everyday practice and experience-based learning (Binder and Lundsgaard, 2014). In this way, the focus becomes directed more towards interconnecting aspects of workplace design and ways of working. Moreover ways of working are changing; the digital (paperless), ways of working are now manifested in altered office design concepts, e.g., activity-based and flex offices (Appel Meulenbroek, 2016; Nag, 2019). These altered office designs are also motivated by reduced office space (in square meters), meaning reduced costs, as well as, in sustainability terms, and reduced energy use.

Recent examples show that design dialogues are fruitful tools for the exploration and promotion of change. This concerns the understanding and modification of systems within organizations, but it also serves a broader educational purpose as a means of learning, educating, and communicating sustainability issues. We will conclude this chapter with two contemporary examples where design dialogues are illuminated in this fashion. First, we will look at the development of a design dialogue kit/method: *Multimodal Origami* (MO) in the research and design project Vis'man.[4] Then we will shortly examine the design dialogues, *Planting Tomorrows* which took place at the Nobel Prize Teacher Summit.[5]

Mulimodal Origami in Vis'man

The Vis'man project was carried out together with participants from four industrial companies and one municipality, all of which were lean-inspired organizations. The aim of the project was to develop new concepts for

---

[3] *The million program* (translated from the Swedish word "miljonprogrammet") refers to the house building policy and industrialized housing project that took place in Sweden between 1965 and 1975. The housing shortage was acute, and politicians decided that one million homes needed to be built within ten years. During this period, apartment buildings, terraced houses, chain houses, and single-family houses were built. But the million program is mainly associated with the large, high-rise buildings in the suburbs.

[4] Vis´man stands for Visual and Spatial Management and Communication from the users' perspective. The project was granted by The Swedish Knowledge Foundation.

[5] The Nobel Prize Teacher Summit is an international teacher conference held in Stockholm every year in October. On the 11th of October, when Planting Tomorrows was on the program, teachers from over 30 countries met Nobel Prize Laureates, top scientists, and peace activists focused on the climate issue, and contributed to workshops. In 2019, "Climate Change Changes Everything" was the overarching theme.

visual management boards (or boards), or alternatives, e.g., digitalized equivalents (Florin and Söderlund, 2018; Söderlund et al., 2020). The purpose was to explore and describe the role of the visual and spatial design elements, their meaning related to the use of these boards from the users' perspective; as well as to identify how visual and spatial design affect information sharing, and how this can be interlinked to a whole information flow within an organization. In lean-inspired organizations, these visual management boards are central to daily planning and communication. Personnel from different parts of the organization gather around the boards for short meetings, with the purpose of updating and discussing production status and flow, and transferring information between different shifts.

Prior to the development of MO, interviews, field studies, a photo elicitation study, and a heuristic evaluation were conducted. These studies produced a vast amount of data in regard to the board's content, placement, and the organizations' use of visual profiles, as well as individuals' use of standard metaphors and symbols. However, this material could not fully capture the role and meaning of visual management boards, nor did it provide insights into the information flow between different boards (and units) and how it related to other visual management devices in the organizations. There was a need to learn more about the information and communication system as a whole, together with the users. This is a short background on why MO was developed, paired with the insight that communication artefacts (such as visual management boards) are embedded in circumstantial aspects that concern the larger context.

Florin and Söderlund propose (2018) that visual management design, and research on visual management, benefit from collaborative methods that undertake an intertwined design thinking and systems thinking approach. This connects to explanations of systems as "functional units or elements relevant to the phenomena under consideration, such as the organs of the body" (Buchanan, 2019). Consequently, these visual management boards must be analyzed as a part of a larger system with its interrelated things: the assembling of the artefact itself, its placement, the users and their understandings and use, as well as the lesser parts such as graphical elements, texts, diagrams, pictures, colors—all interconnected parts in a larger information and communication system.

Ideas behind MO

The primary idea behind the design of MO is the boundary object theory (Star and Greisemer, 1989) earlier presented in the section, "Artefacts in group processes". According to this theory, artefacts are considered to be resonant entities, which are useful in heterogeneous collaborations. In the next layer follows the idea that it is always one's own horizon of understanding that is the prerequisite for interpretation, as the philosopher Hans George Gadamer pronounced in the 1960s (Gadamer, 1960/1997). In other words, one's personal view of the world and its phenomena is the foundation for how and what one can understand. Gadamer suggested that everything is related to context, history, and language, and that nothing can be considered in isolation beyond experience. In spite of this, understanding and communication also have dimensions that relate to our human capability to produce artefacts (specifically pictures), and how we use them in different mediating processes. Understanding, as such, has a dialectic structure, with continuous production of inner pictures piling up in layers inside our heads, in constant dialogue with stimuli from the outer world which are changing over time (Florin, 2019). In this regard, language and conversation are considered to be the primary mediums for understanding, and it is necessary to highlight artefacts as being a part of language in dialogues and negotiations (Florin, 2015).

What MO is—and what it does

Multimodal Origami (MO) can be briefly described as an exploratory, co-creative design dialogue method or tool that enables the recognition of different voices and interests in heterogeneous groups. MO is designed as a board game that can capture how different visual management boards relate to one another, and how the larger system in an organization is impacted by the information flow between the different boards (meetings). MO is based on visual and spatial components and gamification aspects.

The toolkit includes a game board (white cardboard) and tokens: a set of full-body characters (single characters and groups of characters), and abstract things with a more ambiguous design. All of which can be used to represent different things, such as technical devices or other information artefacts. Pencils, markers in different sizes and colors, scissors, small wooden sticks, and oval pieces of wood are also included in the kit. The characters (tokens) are designed in a lo-fi manner, which invites the participants to alter them and take notes directly on them. Together with this beta-version of MO, two sets of characters were developed, and used depending on the type of organization (see Figures 1 and 2).

MO also has a set of specific roles that are assigned to the participants in writing, and also orally presented by the scaffolding game leader. In short, the participants are instructed to choose characters that represent themselves and the staff members who are participating in the specific meetings that they are involved in and relate to. The most central visual management boards(s) and meeting(s) are then marked on the game board with a dashed or solid circle (the participants negotiate on what they find to be the most important board and meeting). The participants

Figure 1. A MO session in the moment that an error in the information flow was detected in the municipal caregiving service organization. It was discovered that two meetings only delivered information to each other which impacts care giving activities, for instance.

Figure 2. A MO session, when the workshop participants point out a specific vulnerability in the communication flow. In this case, one single person was functioning as the information carrier from two important meetings to the core unit at the company.

are also encouraged to take notes, directly on the canvas and tokens (or post-its), with information on where in the organization specific meetings are held, how many participants are attending, how long the meetings are, and how often they are held. Dashed lines represent meetings that receive or send information to other meetings (or information channels), and the solid lines represent the meetings that do not. All meetings are related to the most central board that is first marked on the game board, then all meetings are represented and built around a value centric perspective, stating from the middle—building a map of sorts. Finally, arrows are added to visualize the interactions between meetings, boards, personnel, and other information artefacts and channels. The MO sessions are recorded in order to capture the participants' dialogues and reasoning while playing. The scaffolding game leader takes photographs during the session to capture the game board scenery and how it develops.[6] The outcome is a multimodal set of data: recorded dialogues, photos of the game board scenes with written notes and spatial relationships. Structures and hierarchies within different parts of the organization become visible in a diagrammatic way, expressed from a value centric perspective, and the information flow between boards and their relationship to and within the whole information and communication system becomes visible. Both the discussions and scenery progressions supply rich material for possible analysis.

An interesting outcome is that, in some cases, directly after playing MO, participants from the organizations made adjustments to their organizations directly, triggered by what they themselves observed during the MO sessions. One error related to the information flow in the municipal caregiving service, alerting one that, "two meetings only deliver information to each other", which means a stop in the intended information flow in regard to the planning of care giving activities, for instance (Figure 1).

Another example was when the participants from two manufacturing companies, during a parallel MO session, pointed out that only a single person had responsibility for transferring information from two of the most important meetings (in one of the companies) to the core unit in the organization. This was considered to be a great vulnerability for the organization and, therefore, an adjustment was quickly made by adding personnel to this task (Figure 2).

Planting Tomorrows

The Nobel Prize Teacher Summit is an international teacher conference where teachers from over 30 countries meet and interact. In the program, Nobel Laureates, scientists, and activists contribute with lectures and workshops from different angles and research fields, to highlight and discuss a theme of great importance for education worldwide. In 2019, the theme was "Climate Change Changes Everything". The focus was on how to give important knowledge about climate change while, at the same time, striving to avoid enhancing fear and anxiety in regard to the topic.

The art and design driven design dialogues, called Planting Tomorrows, were designed in this context, with the goal of altering the scenario of climate change.

Ninety of the participating teachers at the Nobel Prize Teacher Summit were invited to interact, experience, learn, and make their imprints in four

---

[6] The scaffolding game leaders in the MO sessions were researchers within innovation and design. One of them also planned and formed the for-purpose-designed artifacts, including tokens, game roles, et cetera.

parallel workshops (nodes and acts), that mirrored important systemic aspects related to climate change, assembled in Planting Tomorrows. The sessions were held on the top floor of Munchen Bryggeriet in Stockholm, in the Riddarsalen venue, a 220 square meter room with a ceiling height of eight meters.

With an introductory talk, delivered in a poetic manner, participants were welcomed and guided to the four parallel workshop nodes. They were told that we now "zoom from a hovering perspective, carried by wings through the clouds, towards our soft globe, and the grounded child" (Florin and Hvistendahl, 2019). The four workshop nodes (acts) were called: Systems, How to Handle Change: a 30 min act on Scenario Mapping, The Human Poem: a 30 min act around the Soft Globe; Cloud Sourcing: a 30 min act of hands-on shaping; and Planting Tomorrows, a 30 min act to make space for futures and planting seeds of tomorrows (and todays) (Ibid., 2019). The naming of these four workshop nodes were a part of the overarching design, and each workshop was designed to supplement the other, with the attempt to map out what boundaries and specifics we need to be aware of and picture—in terms of interconnecting relations among things and people across the globe—in order to be able to educate about climate change. Hence, in a way that avoids enhancing fear and anxiety among young people. Figures 3–5 show examples from three of the nodes.[7]

Figure 3. Participating teachers discussing and building poems that offer commentary on the personal and universal challenges related to climate change in a session called "The Human Poem: a 30 min act around the Soft Globe".

---

[7] All participants in Planting Tomorrows agreed that photos taken during these sessions could be used in future scientific publications and other texts produced by Ulrika Florin and Andrea Hvistendahl.

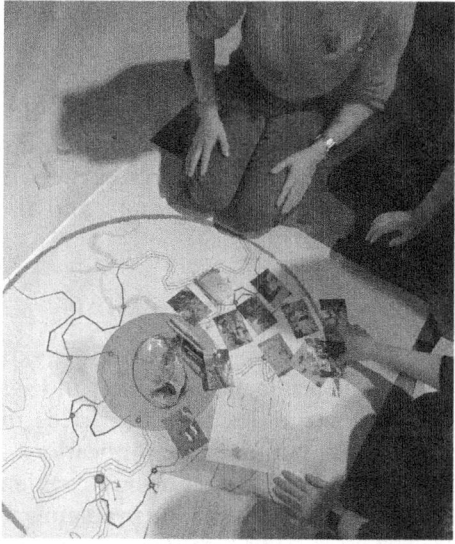

Figure 4. Participating teachers discussing and exploring the themes of Transportation and Infrastructure—and the intersection of natural or man-made systems and the interrelations, in a session called "Systems, How to Handle Change: a 30 min act on Scenario Mapping".

Figure 5. The Grounded Child, the silent moment after the participants were finished with their created rituals on Willingness to Make Sacrifices, in a session called "Planting Tomorrows, a 30 min act to make space for futures and planting seeds of tomorrows (and todays)."

The core idea of Planting Tomorrows, on physical and practical levels, was to recycle and re-assemble art and design objects on related topics, earlier commissioned, used or shown publicly, by Ulrika Florin and Andrea Hvistendahl, in order to build the basis for the workshop with an artistic quality. The idea that permeated the workshop design was that the

126 *Different Perspectives in Design Thinking*

expression forms and designs would be appealing and invite interaction while, at the same time, being legible as a basis for specific thematic design dialogues.

For all four workshop nodes, there were specific instructions and a scaffolding workshop leader (who could be a designer, artist, design researcher, or design teacher) handing out printed instructions and presenting the instructions orally to the groups, and also participating in the dialogues.

Concluding reflections

As stated in the beginning, there are good reasons to explore user insights in more extensive ways. The artefacts, processes, and systems that impact many people are rooted in complex problems. This means that poorly grounded solutions can lead to unforeseen implications—also on systemic levels. In order to grasp how a complex design problem can be defined, and to recognize what boundaries a specific problem spans over, we need ways to picture how systemic, intersecting aspects are related. This is to be able to discuss what sorts of things are, or can be, affected by a suggested design in terms of interconnecting relationships (Florin, 2019). Figure 6 illustrates such relationships.

The area in the middle of the illustration, the eye pupil, zooms in on the specific system in question—the primary boundary, where, e.g., organizations and ecosystems can be defined. This area is also influenced by other interconnected systems, when their paths and nodes enter the defined area and cross the boundary. If we shift attention to the outer circle in the illustration, we find that it is not that obviously defined; there

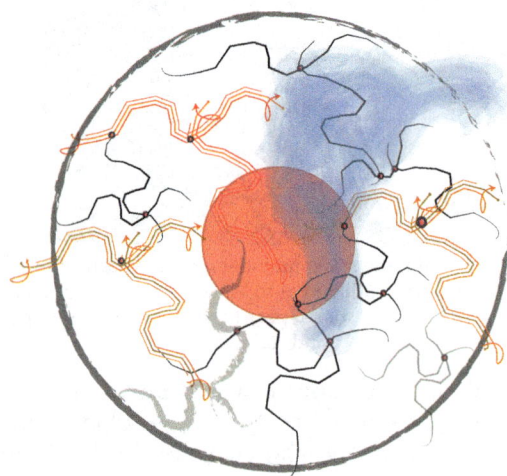

Figure 6. An illustration of how a complex design problem can relate to and be interdependent with other things and systems, with intersecting paths and nodes.

is even a large opening in the rough, uneven outline where a cloud-like formation enters. This cloud-like formation signifies uncertainties and marginal things and/or people outside and inside systems which are either affecting them or affected by them (Ibid., 2019).

Holding design dialogues means having a holistic view on design, and that the modification of things such as spaces, buildings, districts, systems, and processes are in focus. However, it also means that the design can be outlined and decided upon not only with regard to the qualities that it needs to attain, but also with regard to how these qualities can be expressed in a solution. This is achieved through the joint effort of teams that include designers, researchers, stakeholders and users, in series of time-concentrated and content-structured activities (workshops) that aim to develop visions, goals and common understandings—and actually create design proposals, partly together. This is to generate well-grounded and agreeable design. The design dialogues method is useful for sustaining anchored designs that promote a future democratic society. The method combines design thinking and systems thinking and takes different knowledge forms into consideration, which includes tacit, experience-related, and scientific knowledge.

References

Appel-Meulenbroek, R. 2016. Editorial: Modern offices and new ways of working studied in more detail. Journal of Corporate Real Estate, 18(1): 2–3.

Bannon, L.J. and Ehn, P. 2012. Design. pp. 37–73. In: Simonsen, J. and Robertson, T. [eds.]. Routledge International Handbook of Participatory Design (Abingdon: Routledge, 06 aug 2012), accessed 25 June 2020, Routledge Handbooks Online.

Binder, T. and Lundsgaard, C. 2014. Designdialoger om rum og arbejde. Tidskrift for arbejdsliv, 16: 2, 30–45. https://doi.org/10.7146/tfa.v16i2.108963.

Björgvinsson, E., Ehn, P. and Hillgren, P.-A. 2010. Participatory design and 'democratizing innovation'. pp. 41–50. In: Keld Bødker, Tone Bratteteig, Daria Loi and Toni Robertson [eds.]. Proceedings of the 11th Biennial Participatory Design Conference (PDC '10). New York: ACM Press.

Björgvinsson, E., Ehn, P. and Hillgren, P.-A. 2012. Design things and design thinking: Contemporary participatory design challenges. Design Issues, 28(3): 101–116.

Buchanan, R. 1992. Wicked problems in design thinking. The MIT press. Design Issues, 8(2): 5–21.

Buchanan, R. 2019. Systems thinking and design thinking: The search for principles in the World we are making. She Ji: The Journal of Design, Economics, and Innovation, 5(2): 85–104.

Checkland, P. 2000. Soft systems methodology: A thirty year retrospective. Systems Research and Behavioral Science, Syst. Res., 17: 11–58.

Cross, N. 2001. Designerly ways of knowing: Design discipline versus design science. Design Issues, 17(3): 49–55.

Cross, N. 2007. Designerly Ways of Knowing. Birkhäuser Verlag, Basel.

Dewey, J. 1958. Experience and Nature (Rev. ed.). Mineola: Dover Publications, Incorporated. First published 1929.

Eriksson, Y. and Florin, U. 2011. The relationship between a model and a full-size object or building: The perception and interpretation of models. In the Proceedings of ICED

11, 18th International Conference on Engineering Design: Impacting Society Through Engineering Design, pp. 194–203.

Eriksson, Y. 2017. Bildens tysta budskap: Interaktion mellan bild och text (2nd ed.) Studentlitteratur, Stockholm.

Florin, U., Eriksson, Y. and Orre, I. 2012. Designing a model of the unknown: Artistic impact in a chain of skilled decisions, International Design Conference, Design2012. In the Proceedings. Sociotechnical Issues in Design, vol. 3.

Florin, U., Orre, I. and Eriksson, Y. 2013. Collaboration for the improvement of tolerance: Artistic practice in a societal context. The International Journal of Social, Political and Community Agendas in The Arts, accepted 2013 printed 2016.

Florin, U. 2015. Konstnärskap i samspel: om skapande arbetsprocesser i myndighetsledda samverkansprojekt. (English title; Artists in interaction: Creation processes in official collaborative projects.) Doctoral dissertation. Mälardalen University Press Dissertations, No. 179.

Florin, U. and Söderlund, C. 2018. Re-Designing Information Boards: Inter-woven Design Thinking and Doing. Abstract in the Handbook of the OR60 'Anniversary' Conference. The Systems Thinking stream. The Operational Research Society OR. Lancaster University, UK.

Florin, U. 2019. Design Thinking: Bridging Different Understandings and Elaborating in Designerly Ways. Invited talk Friday 29 Nov. 2019, Centre for Systems Studies Research Seminar, University of Hull, Faculty of Business, Law and Politics (unpublished manuscript).

Florin, U. and Hvistendahl, A. 2019. Planting Tomorrows. Workshop Compendium. Nobel Prize Teachers Summit.

Florin, U. and Eriksson, Y. 2020. Visual awareness aiding communication. The International Journal of Visual Design, 14(2): 21–33. doi:10.18848/2325-1581/CGP/v14i02/21-33.

Fröst, P. 2004. Designdialoger i tidiga skeden. Arbetssätt och verktyg för kundengagerad arbetsplatsutformning. (English title; Design Dialogues in Early Phases of Building Projects - methods and tools for customer engaged workplace design.) Doctoral dissertation. Chalmers University of Technology, Göteborg.

Fröst, P., Gustafsson, A., Eriksson, J. and Lindahl, G. 2017. Designdrivna dialoger för arkitektur och samhällsbyggnad. Sweco/Chalmers.

Fröst, P. 2020. Design-driven dialogues for healthcare architecture. pp. 202–212. In: Gromark, S. and Andersson, B. [eds.]. Architecture for Residential Care and Ageing Communities: Spaces for Dwelling and Healthcare, 1st ed. Routledge. https://doi.org/10.4324/9780429342370.

Gadamer, H.-G. 1960/1997. Sanning och metod. (Original title: Warheite und Methode.) In translation by Arne Melberg. Daidalos. Göteborg.

Göransson, B. 1983. Datautvecklingens filosofi: Tyst kunskap och ny teknik. Carlssons. Stockholm.

Hartig, T., Astell-Burt, T., Bergsten, Z., Amcoff, J., Mitchell, R. et al. 2020. Associations between greenspace and mortality vary across contexts of community change: a longitudinal ecological study. J. Epidemiol. Community Health, 74(6): 534–540. doi:10.1136/jech-2019-213443.

Horizon. 2020. https://www.innovationplace.eu/news/multi-actor-approach-essential-in-horizon-2020-work-programme-food-security; https://ec.europa.eu/eip/agriculture/en/news/brochure-"multi-actor-approach" (retrieved 6 March 2021).

Jones, P.H. 2014. Systemic design principles for complex social systems. pp. 91–128. In: Metcalf, G. [ed.]. Social Systems and Design 1. The Translational Systems Science Series. Springer, Japan.

Kang, S. and Tversky, B. 2016. From hands to minds: Gestures promote understanding. Cogn. Research 1, 4. https://doi.org/10.1186/s41235-016-0004-9.

MDH ger världens lärare interaktiva verktyg. 2019. (English title: MDU, Mälardalen University, provides the world teachers with interactive tools) https://www.mdh.

se/artiklar/2019/oktober/mdh-ger-varldens-larare-interaktiva-verktyg (Retrieved 14 March 2021).
Midgely, G. 1992. The sacred and profane in critical systems thinking. Systems Practice, 5: 5–16. https://doi.org/10.1007/BF01060044.
Midgely, G. 2000. Systemic Intervention: Philosophy, Methodology, and Practice, Springer, 10.1007/978-1-4615-4201-8.
Midgely, G. 2014. Research Memorandum, Systemic Intervention. ISBN 978-1-906422-32-5.
Molander, B. 1992. Tacit knowledge and silenced knowledge: Fundamental problems and controversies. pp. 9–31. *In*: Göranzon, B. and Florin, M. [eds.]. Skill and Education: Reflection and Experience. Artificial Intelligence and Society. Springer, London. https://doi.org/10.1007/978-1-4471-1983-8_.
Molander, B. 1996. Kunskap i handling. Daidalos. Göteborg.
Nag, P.K. 2019. The concept of office and office space. In Office Buildings: Health, Safety and Environment. Singapore, Springer Singapore, pp. 3–27.
Nobel Prize Teacher Summit. https://www.nobelprize.org/education-network-nobel-prize-teacher-summit/, retrieved, 14 March 2021.
Olsson, B.K. and Florin, U. 2011. Idea exchange and shared understanding. Design Principles and Practices: An International Journal, 5(4). The Design Collection, CGP.
Papanek, V. 1988. The future isn't what it used to be. Design Issues, 5(1): 4–17. JSTOR.
Polanyi, M. 1966/2019. The Tacit Dimension. Double day and Company, INC, New York.
Pourdehnad, J., Wexler, E.R. and Wilson, D.V. 2011. Integrating systems thinking and design thinking. The Systems Thinker, 22(9).
Sanders, E.B.N. and Stappers, P.J. 2008. Co-creation and the new landscapes of design. Co-Design, 4: 1, 5–18. DOI: 10.1080/15710880701875068.
Star, S.L. and Griesemer, J.R. 1989. Institutional ecology "Translations and Boundary objects: Amateurs and Professionals in Berkley's Museum of Vertebrate Zoology". Social Studies of Science, 19(3): 387–420. Sage.
Söderlund, C., Florin, U., Lundin, J. and Uggla, K. 2020. Participatory involvement and multi theoretical perspectives in visual management design. pp. 1541–1550. *In*: Dorian Marjanovi, Mario Štorga, Stanko Škec and Tomislav Martinec [eds.]. Proceedings of the International Conference on Engineering Design 1, 1541–1550. Cambridge University Press. https://doi.org/10.1017/dsd.2020.46.
Tien, J.M. and Berg, D. 2003. A case for service systems engineering. Journal of Systems Science and Systems Engineering, 12: 13–38.

Recycled Art and Design objects in Planting Tomorrows

1. "The Clown Girl" or "The Intersection" by Ulrika Florin Permanent public art, commissioned by Nacka Municipality 2000.

    https://www.nacka.se/uppleva--gora/kultur-och-museer/konst2/offentlig-konst-i-nacka/offentlig-konst-i-alta/alta-skola-clown-flickan/.

2. "Lanterns in three contexts" by Ulrika Florin 2006, temporary public art commissioned by: Eskilstuna Art Museum, Sörmland County Council, Svenska Kyrkan (The Church of Sweden) and Nyköping municipality. The cultural festival "Different" in Nyköping, 2. Eskilstuna Art Museum's entrance (Fristadsnatta), 3. The dance festival "2", a collaboration project (art and dance) together with Daniel Andersson (dancer); performed at The Nyköping State Theater and The Eskilstuna State Theater.

3. "Interconnectedness and Learning loops" by Ulrika Florin 2019, for BRIGES Sweden.
4. "Falling or Flying" by Andrea Hvistendahl and Esmé Alexander 2001, Kulturhuset Stockholm and Erik Eriksson exhibition hall Stockholm.
5. "Paths of Change" by Andrea Hvistendahl 2017, Ronneby Art hall 2016, Norrtälje Art hall.
6. "The Cloud" by Andrea Hvistendahl 2019 commission by Norrtälje Art hall.

Figures

Figs. 1–5 Photo, Ulrika Florin, 2018

Fig. 6 Illustration, Ulrika Florin, 2019

CHAPTER-8

Design Thinking and Designerly Ways of Knowing in Operational Research Practice

Christina J Phillips

Introduction

Let us begin by understanding what is meant by operational research; according to The Operational Research Society in the United Kingdom *"Operational research (OR) is a scientific approach to the solution of problems in the management of complex systems that enables decision makers to make better decisions."* It is often used to solve messy and complex problems in a wide variety of domains including business and industry, but also government and the military. Operational research uses many different techniques, from those that help to structure the problem domain (working in mainly social settings) to analytics and data science, mathematical/statistical modelling and optimisations.

Some argue that operations research has become differentiated from operational research due to a divergence in the discipline (covered in a later section). Where operations research is more theory based, operational research is more practice based. However, the two terms continue to be used interchangeably by many scholars and both terms are often used to refer to either type of premise. For the purposes of this chapter, we will use the term operational research, and we discuss the divergence of the disciplines in more detail as this is of interest when looking at the discipline from a design perspective.

The use of design in Operational Research and Management Science (ORMS) is undisputed (Keys, 2000), but it is also under researched and often ignored. Through research on systems, the connection between

---

Liverpool Business School, Email: c.j.phillips@ljmu.ac.uk

operational research and design as disciplines can be traced back to the Second World War and the inception of new service provision in the National Health Service in the UK (McDonald et al., 1974; O'Keefe, 1995). Early operational research was all about practical problem solving and rooted in relevance but by the 1960's the push to increase the scientific rigor of operational research began a drift into axiomatic research, which sought to grow knowledge from theory. This fits the culture of science as described by Nigel Cross (1982) but the earlier, solution focused, systems approaches of operational research's foundations look more like his description of a culture of design.

The need for operational research to be put into practice has meant that the links to design have become more profound even as they have become less celebrated. Both the Operational Research Society (ORS), in the UK, and the Institute for Operations Research and the Management Sciences (INFORMS), in the USA, have dubbed operational research as 'The Science of Better' (http://www.scienceofbetter.org/) to emphasise the fundamentally practical, solution and improvement focus of operational research. The axiomatic, mainly mathematical modelling type of research has led to the development of much improved algorithms and methods but the initial scoping of what needs to be improved requires a more designerly way of thinking (Cross, 1982).

There have been calls within the operational research community to seek improved models with a better situational fit by linking more with practice (Sodhi and Tang, 2008), as this will often lead to novel insights and new axiomatic design, as well as signalling practical relevance. Some have started to make the link with design as a way of going about this (O'Keefe, 2014), although they generally err on the use of design science as a tool, as do many in the closely related fields of information systems (Hevner et al., 2004) and operations management (van Aken et al., 2016).

Design thinking in operational research

Most operational research papers using the word 'design' are using it in an improvement sense, and the operational research community rarely study, or write about, how they are applying design thinking. In this respect they are using and developing 'designs in/for practice', as opposed to using 'design as practice' (Kimbell, 2012). In almost all cases 'design' is used to infer an improved offering, with the artefact often a mathematically derived decision model or algorithmic solution method, but without methods that also help designs to work with human systems, the improved mathematical solutions can remain unused.

Creating an improved artefact, which is appropriate to the situation, via a sympathetic process of pattern synthesis, fits some of the definition of a designer put forward by Nigel Cross (Cross, 1982). As an example of this occurring in operational research let us look at a classic case of using hard operational research (highly structured and bounded) to derive a new optimisation model for an organisational/industrial problem. To begin with the problem will need to be well defined for an optimisation algorithm to be derived, the data has to be clean and available, and the parameters must be clear. Sometimes the solutions to these problems take so long to solve in a computer that it is infeasible (known as NP-hard), and at this point the operational researcher might try to reduce the problem by 'trying out' different rules of thumb, or heuristics. Do they do this on a purely scientific/analytic basis or do they use their existing knowledge, and that of their clients, to make informed guesses and 'see what works', as a designer would? If the rules of thumb that the design uses resonate with the processes and understanding of the human user, they are more likely to understand the output and use it.

Frequently it is the case that a beautifully designed algorithm such as this will have to be used by a human to augment their decisions. The use occurs within a kernel of activity that includes the human user's knowledge, and physical engagement/perception of the algorithm and the accompanying user interface, and these happen within the systems and structures in which they perform their working practice. No matter how well the algorithm works, if the interface and interactions fail, the design will too.

If we look at this process through the practice lens used by Lucy Kimbell, in her re-thinking of design thinking (Kimbell, 2012), we see the design element of operational research coming about most prominently when dealing with socio-technical systems, joining the solution methods which operational research provides with human actors. Kimbell (2012) points to the need to be aware of how to think like a designer at these moments, new normative routines and structures have to be grown at the locus of the material, cognitive, knowledge, process, agency and existing structure.

Designers know about artefacts (Cross, 2001; Simon, 1997) and so do operational research practitioners and academics, the designing lies in the moments where creative moves are made to adapt or extend the known artefacts/patterns. This is informed by knowledge gained through repeated practice and creation, which is then reflected upon, and is in turn influenced by the problem owners, this reflective decentred practice of operational research development would fit a description of design thinking (Cross, 2001).

For the rest of this chapter we will use the practice approach put forward by Kimbell (2012) to see where operational research relies on design thinking at the locus of practice, even when it is not directly recognising it as such. Design thinking is a decentred activity that occurs at a socio-technical interface and which requires reflection and discursive opportunities. Sometimes this comes in the form of specific methods designed to prompt this type of thinking such as persona creation and possible futures. At other times the use of these ideas is implicit within the operational research methodologies used, in particular in the realm of Soft Operational Research and Problem Structuring (SORPS), which we look at in more detail. In some cases, a design stance resides within the philosophical underpinnings such as in the case of management cybernetics or systems thinking. The design thinking as practice lens will allow us to discern how much design thinking is embedded within these different streams of operational research.

We also turn to the definition of 'designerly ways of knowing' as laid out by Cross (1982);

- Designers tackle 'ill-defined' problems.
- Their mode of problem-solving is 'solution-focused'.
- Their mode of thinking is 'constructive'.
- They use 'codes' that translate abstract requirements into concrete objects.
- They use these codes to both 'read' and 'write' in 'object languages'.

How do these ways of knowing manifest in operational research practice, explicitly or otherwise?

Herbert Simon and the science of design

The growth in the application of science and mathematics to problems of war and management led to the development of operational research at the same time as it led to a desire to make design practice more analytical, to create a 'science of design' and a 'design science' (Cross, 2001). Many of the techniques of operational research fed into methods of design, as called for in the 1960's by the systems thinker and inventor of the geodesic dome Buckminster Fuller, and later expanded by Herbert Simon in his 'Sciences of the Artificial' (Cross, 2007; Simon, 1997).

It is worth expanding upon the influence of Herbert Simon at this point as he was a highly influential scholar in design, economics, management science, computer science and operational research. Although his original training was in cognitive psychology, his PhD applied this in administration, and he became interested in how people make decisions, this prompted him to start formulating his ideas on bounded rationality (for which he later received a Nobel in economics). He taught across management disciplines, but also in computer science and psychology. Simon was also one of the early proponents of Artificial Intelligence (AI) and received the Turing Award for this work along with his collaborator Allen Newell. It was in this domain that he wrote on ideas such as the definition between ill structured and well-structured problems. He argued that the boundary between them is vague and fluid but that, in general, ill structured problems are made up of well-structured sub problems that are evoked and adjusted at different stages of the design process. This digs at an underlying justification in design for making many alternatives, in order to see what works, but that also has the best fit with the desired end goals, both systemically and locally.

Simon's work on bounded rationality and satisficing solutions in developing an understanding of the artefacts we create can be recognised by designers, engineers and operational researchers alike. We can use these concepts to help us manage, replicate, and achieve better representation for decision making; *"prospective artificial objects having desired properties are the central objective of engineering activity and skill. The engineer, and more generally the designer, is concerned with how things ought to be how they **ought to be*** (his emphasis and repetition) *in order to attain goals"* (Simon, 1997) This chimes with 'the science of better' descriptor of operational research and the generic goal of the designer to work toward an improved end design.

Herbert Simon coined the term *'the science of design'* as *'creating the artificial'* and posited that *'Everyone designs who devises courses of action aimed at changing existing situations into preferred ones'* (Simon, 1997). He lamented the loss of design teaching on professional courses and felt it damaged competence, this has been remedied in many areas (i.e., engineering and medicine) but it is still rare to receive design training in management or operational research courses, yet it was these that he invoked when talking about a rigorous science of design. The correctness of his arguments regarding rigour in design have been argued (Cross, 2001, 2007), and the requirement for precise definitions and defined logic run contrary to the fluidity of definition for design thinking (Kimbell, 2012). We need both a science of design–design science, AND a paradigm of design–design thinking, in order to be able to invoke methods of design and deal with them in designerly ways (Cross, 2001). This is true in operational research as much as it is in engineering design, or architecture.

Over the next few sections we take a look at many of the cross disciplinary techniques, frameworks and methodologies that are used in both operational research and design, and we will also look at some of the personalities who were highly influential in both disciplines. However, this book is about design thinking and it is that aspect of these various cross overs on which we will maintain our focus.

Soft Operational Research and Problem Structuring (SORPS) and the founders of operational research

There are many who should be included as the founders of operational research, but we will take a look at just two of these proponents whose influence extended from the USA to Europe; C. West Churchman and

Russell L. Ackoff, who together, authored one of the early seminal texts of operational research with Leonard E. Arnoff (Churchman et al., 1957). Both Churchman and Ackoff were rooted in ideas of whole systems design and Churchman was particularly keen on ethics. Churchman was the founding Editor in Chief of the journal *Management Science*, which is viewed as one of the top Operations Research and Management Science (ORMS) journals, but in the late 60's ORMS moved away from design and systems science to focus more on the development of models and axiomatic solutions to defined problems.

By the 1970's Churchman was writing books on design (Churchman, 1971) and systems (Churchman, 1979), while Ackoff had become an ORMS apostate (Ackoff, 1979), declaring it to be dead in the water! They both found that the emphasis on novel mathematically derived, generalizable models was pushing operational research away from the very clients it was originally envisaged to help. Ackoff moved away from the USA's highly mathematically derived way of doing operations research, and found a new home publishing frequently in the UK's *Journal of the Operational Research Society*, later moving into design research (Ackoff et al., 2006; Kirby, 2003). The momentum he evoked with this historic move gave birth to a renewed focus on techniques that resonate with design thinking, and to this day one of the distinguishing features of operational research in the UK, is its dual focus on social structuring techniques alongside primarily mathematical methods.

Although it has been a rocky journey (Ackermann, 2012; Mingers, 2011; Ranyard et al., 2015) by defining a separate stream with in the discipline, Soft Operational Research and Problem Structuring (SORPS) has created and developed many techniques to deal with the problems associated with putting operational research into practice. These have in turn spread beyond operational research to be used in design, computer science and engineering as well as many other disciplines. Design and SORPS are closely related (O'Keefe, 1995; Royston, 2013) but often are not using design as practice, as would befit design thinking, but rather stem from the analytic school (O'Keefe, 1995) and are performed as designs in practice (Kimbell, 2012). Design thinking is not implicit within these methods, although many pay attention to the same dimensions, i.e., ethics, participation, systems thinking, discursive practice and reflection, not all do, and there is still a tendency to put the modeller/facilitator into the centre of the practice. Decentred iterative and reflective design is not necessarily embedded in the SORPS methodologies, and appears to come from individual practitioners preferred stance, with the exception of Soft Systems Methodology (SSM) (Checkland, 2000; O'Keefe, 1995).

Soft Operational Research, Problem Structuring methods and their designerly ways

We now take a look at some of the better known and used methods of SORPS (Soft Operational Research and Problem Structuring) to give an overview of the techniques that have grown out of its practice.[1]

SODA

Strategic Options Development Analysis (Ackermann and Eden, 1989) is a complete methodology for bringing together differing opinions to facilitate group decisions and negotiations. There is an underlying process of exploring the individual cognitive stance through cognitive mapping (Eden, 1988), this form of sophisticated 'sketch' of an individual's cognitive perspective comes from the work of George Kelly and his work on Personal Construct Theory, sometimes called Personal Construct Psychology.

The idea is that we each have a construction of reality, which allows us to make sense of what is happening to us and informs our decisions as to how we will react. Kelly developed the Repertory grid, Eden extended this to mapped constructs and dubbed them cognitive maps (Eden, 1988), these could then be bought together to develop causal maps of the organisational focus, prompting discussion, understanding and shared commonality. Eden suggests attending to 3 key aspects that each follow from the other (corollaries) in one's thinking when working with individuals and teams;

- Individuality
- Sociality
- Commonality

So, each individual will have their own version of events (individuality) because everyone has their own way of thinking about (constructing) the world around them. When we work together we make some headway into understanding another's way of constructing the world and we may influence that construction (sociality). When a team works well together it is because the members, to some extent, understand the constructions of others and how they might influence it. Thus they are able to communicate using the same, or similar constructs, which prompts easier

---

[1] Some of these techniques may be familiar, and it may come as a surprise that their foundations were in operational research. For a full and in depth coverage we would refer the reader to the seminal text by Jonathon Rosenhead, Rational Analysis for a Problematic World, or the more recent updated version Rational Analysis for a Problematic World Revisited: Problem Structuring Methods for Complexity, Uncertainty and Conflict 2nd Edition, Rosenhead & Mingers.

and more comprehensive communication (commonality). The cognitive map is an artefact designed to let others gain insight into another's way of constructing the world such that they can mutually envisage an improved future state and agree upon actions to create it (Eden, 1988).

In an attempt to map out the constructs of each individual, the cognitive map asks one to provide both an affirmative and a negative response; I think this ... *rather than* this.... This is based on the conversion of Kelly's ideas into a format that works well with client situations, as Eden, 1988 put it;

> *"Firstly, man makes sense of his world through contrast and similarity, that is meaning in the context of action derives from relativism. Secondly, man seeks to explain his world—why it is as it is, what made it so. And thirdly, man seeks to understand the significance of his world by organising concepts hierarchically so that some constructs are superordinate to others."*

Thus the cognitive map starts with a theme, for instance, design thinking, we then write down the constructs, as an individual. So we might sketch the idea that 'cognitive mapping can foster design thinking', so we write down all of those 'things'—the constructs—which we think enable this to happen and also which, we think, stop this from happening or are incompatible with design thinking. We then arrange these hierarchically and link them with positive or negative signs, which allows us to see if they foster, or negate, design thinking to our way of thinking—our personal construction. They provide a 'map' of how one thinks about the subject and what corollary's one is looking for when exploring a particular subject, i.e., I think 'iterations are part of design thinking' and 'cognitive maps can be created in an iterative way', therefore 'the ability to alter my cognitive map until I am satisfied' must follow. The cognitive map is a coded artefact, which has its own unique object language, thus producing a cognitive map is a distinctly designerly process.

The SODA (Strategic Options Development Analysis) methodology takes this a step further by creating these maps at an individual level and then bringing them to each of the other parties so that they can gain a view of the constructs of others and the strategic options that might be available. The final step is to bring everyone together in a workshop to hammer out the final options and create a causal map for the situation as a whole.

We can see that not all of this methodology fits with design thinking, as it has strong boundedness of method, which is to help provide testable quantitative data for analysis—a more analytic approach. However, the cognitive mapping process is highly discursive, using participation and multiple iterations in order to hone down thinking to a few clear constructs. The end result and iterative practice relies on a sketch, the SODA process is open to further iterations and is also broad bounded enough to encourage discourse that fosters reflexivity on the part of participants.

In what we might call a designerly way the founders of the SODA (Strategic Options Development Analysis) method responded to observations of clients using SODA by creating JOURNEY Making (JOintly Understanding Reflecting and NEgotiating strategY) which took the process forward by creating a decision model which could be used to create agreed defined goals and processes. One of the aspects in these methods emphasised by the founders, is the use of software to help create definition and provide repeatable and comparable exercises by recording qualitative data in a defined way. More of a design in practice than design as practice, with the boundaries providing a basis for negotiating complex and messy problems in a bounded way that forces definition to emerge.

About the founders of these methods: Fran Ackermann is now a Research Professor at Curtin School of Management, Perth, Australia. Her work focusses on complex decision making and she is *'keen to ensure that strategy is developed that is both robust and owned'*. Colin Eden's initial training was in Mathematics but he found mathematical techniques were not enough when used alone to try to solve engineering problems in industry, this led to his interest in action research and to the development of cognitive mapping.

Together Colin and Fran have been highly influential in the robust development of decision support systems and group negotiation techniques, and they have worked in many different organisational domains. Their work continues to inform the development of Problem Structuring Methods in Operational Research and has been instrumental in solving numerous problems outside of the academic domain.

Action research

Eden and Ackermann developed both SODA and JOURNEY making using action research over multiple interventions. Action research cannot be claimed as an operational research method but it has been used extensively in operational research practice and in the closely related subject of information systems (Baskerville and Wood-Harper, 1996; Mumford, 2006). Many problem structuring methods are said to be action research techniques, or have been grown out of Research Oriented Action Research (ROAR) (Eden and Ackermann, 2018) over many interventions, and some important papers on using action research have been written by operational researchers (Checkland and Holwell, 1998; Eden and Ackermann, 2018; Eden and Huxham, 2006). This is not surprising when we think of the socio-technical aspects of operational research, where we are often trying to define and structure a problem such that it might become amenable to modelling and analysis. To do so in a way that is sympathetic to the human end users precipitates the use of reflexive methods such as action research.

In action research one observes a situation, working with stakeholders in a decentred and reflexive fashion, the facilitator/analyst should not take the expert mode or impose normative values, as a team you work toward a proposed intervention, then observe as the intervention is put into practice, give time for reflection then iterate. A good outcome is desirable for both the academic and the participants (if it involves both), and this can be achieved by using a well-defined epistemic lens and accompanying ontology, for instance intervention theory (Argyris, 1970).[2] In this way your personal recording will be bounded within this lens providing recoverable information that could be compared across cases. Triangulated information is essential to improve the rigour of observations; personal notes, others notes, audio recordings, data and more.

Action research is a study technique that provides an opportunity for design thinking as it is participative, iterative and discursive. The facilitator should not be the centre of the action, so it is a non-expert mode, and one must be careful not to impose normative assumptions, so it is exploratory and encouraging of creativity in the potential solutions from all of the participants. Perhaps this is in part why action research, and associated methodologies are popular as techniques to cross the socio-technical boundary. Sketching is not explicitly called for in action research practice but it is often there in one form or another, be that literal sketches or multiple stick it notes and connections. New learning and improved solutions are thought to be positive outcomes from any action research, see these varied examples of using it in operational research and management science (Caniato et al., 2011; Phillips and Nikolopoulos, 2019; Senge and Sterman, 1992).

Often cited alongside action research is design science (Baskerville and Baskerville, 2017; Hevner et al., 2004; O'Keefe, 2014; Royston, 2013). We will not go into detail since we are trying to find the connection with design thinking, not a design method (Cross, 2001) but if you read around this subject you will find that it is often the preferred way of using design techniques in operational research and management science. The method is similar to action research in that it takes a design thinking stance as discussed above, however there is a requirement to produce an artefact by the end of the process, and to some extent the success and rigour of the intervention is judged on the efficacy in use of the designed process or product and the contribution made to new knowledge, i.e., theory.

---

[2] Intervention theory was devised by the organisation scientist and management scholar Chris Argyris. It states that a good intervention should create valid information that is agreed to by all parties, foster the free will of the participants, and gain internal commitment by all those involved in the intervention.

Soft Systems Methodology (SSM)

Another Problem Structuring Method (PSM) that identifies with action research is SSM developed by Peter Checkland (Checkland, 2000). SSM consists a suite of methodologies to help draw out the social and political aspects of a problem domain as well as structuring the problem to make it more amenable to solution. It begins with the premise that in any social situation there is unlikely to be consensus about the solution and although some opinions may change, some may not and that ambiguity is to be embraced. Checkland's ideas are rooted in action research and the appreciative systems of Geoffrey Vickers (Checkland, 1985).

At the very heart of the soft systems methodology lies the purposeful activity, a transformation process which is drawn out through a step called CATWOE and which is often accompanied by a sketching device called rich pictures. In Checklands own words;

"C ('customers') Who would be victims or beneficiaries of this system were it to exist?

A ('actors') Who would carry out the activities of this system?

T ('transformation process') What input is transformed into what output by this system?

W ('Weltanschauung') What image of the world makes this system meaningful?

0 ('owner') Who could abolish this system?

E ('environmental constraints') What external constraints does this system take as given?" (Checkland, 1985)

Rich pictures are used to record these elements to produce 'root definitions' that sketch out the individual/group conceptual models of the activity system. The *Weltanschauung* is often the most used stance in this process as it is explicitly looking for the differing 'worldviews' of the various actors and the acceptance that they may remain different. There are more parts to the soft systems methodology process and for a deeper treatment of the subject we would recommend reading Checkland's own 30 year retrospective (Checkland, 2000). It is used across many different domains today and is claimed as action research, an operational research problem structuring method, it is extensively used in engineering, and is thought by many to be a systems engineering/thinking technique.

Soft systems methodology's ubiquity and popularity across disciplines stems in part from a strongly design oriented practice and also from the highly flexible suite of methodologies which can be cherry picked to suit the circumstances and used in an iterative and recursive way. The process is completely participative when used as an explicit and deliberate methodology, but there is also a second mode of use as a framing device to understand the situation at hand and to interpret evidence. There is evidence both past and present to suggest that the methodologies are cherry picked and used as separate parts outside of the whole to develop understanding rather than to bring about change (Mingers and Taylor, 1992).

The very flexibility of Soft Systems Methodology (SSM) is part of its strength, it has a sophisticated code of its own and uses sketches and models as facilitation objects to prompt creativity and stimulate discourse. The process allows for reflection and iterations of design, both recursively and as a whole. O'Keefe felt that of all the soft operational research and problem structuring methods SSM is perhaps the most designerly in its approach (O'Keefe, 1995).

Once again we see in the background of Peter Checkland, years spent in practice and a multidisciplinary approach to problem solving. After gaining a first from Oxford in Chemical Engineering, Checkland spent 15 years at Imperial Chemical Industries (ICI) working his way up to a senior managerial position before moving across into academia. He created a research programme that was embedded in a Department of Systems Engineering, which used action research, bringing practice and academics together to study and resolve issues in messy and wicked problem spaces (those which can be interpreted in many ways and often change). They partnered with industry to work toward improved situations, if not complete solutions, by identifying meaningful activities for change through active dialogue and consideration for the worldviews of others. The work was rooted in a critique of the highly structured approach to systems engineering of the time. The engaged Masters Programme and subsequent case studies led to the development of Soft Systems Methodology (SSM).

Systems thinking

Within operational research, systems thinking is differentiated from soft operational research and problem structuring, however systems techniques are often used by their practitioners and many consider soft systems methodology to be part of systems thinking.

Systems, operational research and design as disciplines were highly connected in the early 20th century, but over time have become defined in their own right as theory and practice have created separate ontologies and epistemologies (Buchanan, 2019; Kuhn, 1962). However some of the leading proponents of systems thinking, critical systems and systemic problem structuring come from operational research departments and write regularly for operational research journals (Checkland, 1999; Churchman, 1979; Jackson, 1993, 2001, 2009; Jackson and Keys, 1984; Midgley et al., 2013). The links with design thinking are in the messy problem spaces involving multiple viewpoints, which may have ambiguous outcomes, and in the operational research proponents of design thinking such as Ackoff and Collopy (Buchanan, 2019). The links between design and systems are deep and multifarious and beyond the scope of this text, but the design thinking part runs implicitly through much of systems research they are bound at the root as they grew together in the beginning along with operational research.

As Midgley et al. (2013), have pointed out, problem structuring methods can be used as explicitly systemic methods and generally view problems in a systemic way. Recently scholars from backgrounds in operational research have been brought together with designers and engineers to work on the big questions surrounding healthcare systems design (Ciccone et al., 2020). The Healthcare Systems Design Group includes operational research scholars who have worked with real problems in healthcare service provision (Royston et al., 2016), systems redesign using participative simulation (Kotiadis et al., 2013; Tako and Kotiadis, 2015) and pharmaceutical supply chain improvement using human centred design methods (Phillips and Nikolopoulos, 2019).

Systems methods continue to be used in operational research, as was surfaced in a survey for the Operational Research Society, Problem Structuring Methods Special Interest Group, (PSM SIG) see Fig 5. Methods of high use amongst the members were in particular soft systems methodology and viable systems modelling developed by Stafford Beer, a thought leader in the cybernetics movement and the developer of ideas that became management cybernetics, which we cover in a later section.

Strategic Choice Approach (SCA)

"a set of methods to facilitate group communication about complex decision problems" (Friend, 1992). The methods were developed by a group of operational researcher practitioners working with *'messy'* and sometimes *'wicked' problems* (Buchanan, 1992) over many years (Friend and Hickling, 2012; Friend and Jessop, 1976) at the Coventry branch of the Institute for Operational Research (IOR). Strategic choice approach was a response to many years of working with government and research councils on policy and planning decisions. Initially described by John Friend and Allen Hickling, it was developed by a group of operational researchers from the OR Society and social sciences researchers from the Tavistock Institute who worked jointly under the banner of the Institute for Operational Research. The result was a methodology that explicitly set out to aid the planned management of uncertainty (Friend, 1992).

Strategic choice approach is often used within an action research context and was grown primarily through practical application with theory developed over time leading to adoption by the academic operational research community as a soft operational research and problem

144 Different Perspectives in Design Thinking

structuring method. The focus on decisions is what differentiates strategic choice approach from some of the other approaches we have discussed (Friend, 1992) and it was developed primarily within a planning context, in particular, complex situations requiring difficult planning decisions. It creates a locus of activity around 'the planned management of uncertainty' (Friend, 1992), which requires thought around which decisions should be made first and acknowledges the complex web of interdependencies that can come with each decision.

The approach is interactive and iterative, and the use of notes facilitates what Friend calls a *'group memory'*. This is named the *'Shaping'* phase when the problem is given structure by defining the options in terms of uncertainty, see Figure 1.

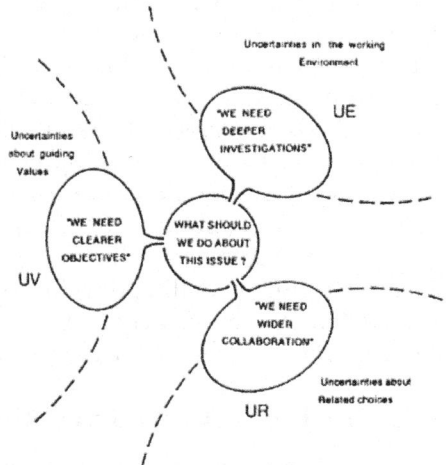

Figure 1. Three types of uncertainty, from (Friend, 1992).

The framework defines areas of thinking about the problem and its attendant decisions and it has phases of development, see Figure 2. The use of this framework provides a codified sketch of the problem and the possible decisions that can provide solution—strategic choices. This process of a codified artefacts creation is very designerly and the iterations of design, high participation, non-expert stance and messy problem domain are resonant of design thinking.

The design phase in the software version consists of defined ways to build decisions, by eliciting assumptions and risks, to build decision schema. This is a design process in practice more than design as practice and provides an analytical means to design solutions. Not so much design thinking as design methods. SCA remains to this day a highly popular technique in planning and strategy and is used regularly by Soft

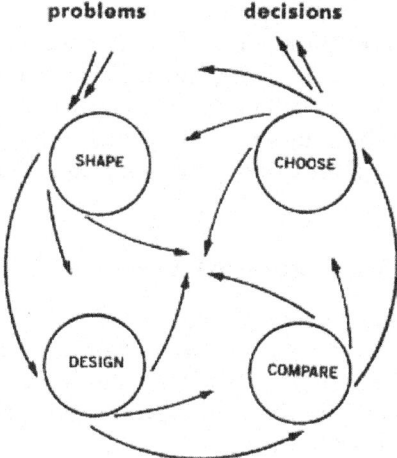

Figure 2. The phases of Strategic Choice Approach (SCA) (Friend, 1992).

Operational Research and Problem Structuring (SORPS) practitioners, often alongside other methods.

We have touched on a few of the main methods, which are used regularly among SORPS practitioners and academics. The *European Journal of Operational Research* has a section dedicated to problem structuring method's (PSM section), which publishes related work that pushes forward framework and methodology design as well as new theoretical developments. This emphasises the importance of these methods to operational research as a whole, the separation of soft methods from hard can be seen by some to be problematic, but it is necessary to drive forward the sciences of soft operational research and problem structuring. Although papers are occasionally published which mix methods, these can be domain cutting in operational research, especially when they address socio-technical problems. Unfortunately, the soft aspects of operational research development and design are often ignored or left implicit. This indicates that teams of researchers with expertise in the different domains are not working together, as they did at the Institute for Operational Research, and this collaboration across paradigm domains should be encouraged.

Soft Operational Research and Problem Structuring (SORPS) links with some of the other areas, which we now discuss, but it has been mentioned first as it is the arena of operational research where most of the design thinking takes place. It is clear that most of the SORPS methods to a greater or lesser extent foster design thinking as they acknowledge a decentred non-expert role for the facilitator/analyst, encourage iterations of development and high participation. The tools and methods foster the codification of problems into a defined structure, or model in a designerly way, and they are used to address complex and messy problem spaces with many competing stakeholder perceptions. There is a tension between the need for bounded codified design, provided by artefacts and tools, and the need to apply design thinking to gain inclusive and effectively structured problem definitions, by allowing ambiguity, multiple perspectives and emergence of ideas.

Management cybernetics

Cybernetics is said to go beyond design by some (Pickering, 2009) but this takes a narrow view of design as always working toward an improved end artefact. Codification of artefacts is conceived a designerly way (Cross, 1982) but design thinking goes beyond end results and goal driven pursuits (Kimbell, 2011). Cybernetics is the study of things as they evolve in time; as Beer saw it, a hylozoist dance between matter, moment and evolution, where matter itself has agency (Pickering, 2009). This is not so far from the design in practice view of Kimbell (2012), which occurs at the nexus of design as it is put into practice, and/or used as practice with humans, processes and materials. If we take this view then there is a strong link between cybernetics and design thinking.

There has emerged an entire field of design dubbed *'Design Cybernetics'* (Fischer and Herr, 2019) and even one of the early proponents of cybernetics drew a link between cybernetics and architectural design (Pask, 1969). There have also been those who draw attention to the very process of iterative and recursive modelling in design as being distinctly cybernetic in nature (Maier et al., 2012).

Cybernetics was first coined by the mathematician Norbert Weiner shortly before the end of World War 2, it is taken from the Greek word kybernetes, meaning steersman. He defined it due to what he saw as a deficit in the modern scientific method which did not address issues of emergence and feedback and which sought to define a complex system by measuring its parts (Krippendorff, 2007; Pickering, 2002). This led to the *'Law of Requisite Variety'* (Ashby, 1956), which (simply put) states that to model/control a system completely the model/control system must contain the same variety as the system it seeks to represent/control. This in turn led to the idea of *'black boxes'*, parts of systems which, when they acted as a whole, were self-regulating until acted upon by some outside input, Ashby called these homeostats, as they were capable of achieving and maintaining homeostasis (balanced dynamic behaviour). In this way the requisite variety is reduced as one only needs to know the simple coupling parameters of the black boxes.

Weiner and others developed an ontology and mathematics to describe a distinctly different epistemology, one that included un-knowable elements, feedback, simple linkages leading to complexity (i.e., automata) and recursive application of systemic behaviour. So, for instance, over a 30 year period Beer developed Viable Systems Modelling (VSM) based on observations of both biological and management systems that achieved homeostasis. In his quest to find *'how systems are viable'* (capable of independent existence) he posited the idea that all viable working systems should follow some of the same rules of regulation and control (Beer, 1984). There is much more to cybernetics but it is beyond the scope of this text, if the reader is interested we would recommend Krippendorff (2007), Pickering (2009) and Beer (1984) as a starting point. Here we are looking at operational research in particular and so now describe the viable systems model and Beer's systems modelling ideas in more detail to see how they reflect what we would call design thinking.

Beer was very particular in the way he defined the viable systems model, it was not, as many argued an analogy to the nervous system, it was an observation over many years of refinement that any viable system had to obey certain rules and structures to become and remain viable. In this respect he designed Viable Systems Modelling (VSM) through observation, feedback, iterations of design and recursive application of ideas that worked. A distinctly design like approach. The first iteration was his observation that the running of a steel mill was an organic system, this was then iterated upon in *'Brain of the Firm'*, improved via his work in Chile (Medina, 2006) and iterated upon in his later book the *'Heart of the Enterprise'* (Beer, 1984).

One could describe this reflexive, iterative, creative, pragmatic and sympathetic evolution as using a design thinking approach, even if Beer did not recognise it as such at the time. He saw it as a process of refinement of comparative analytic/conceptual modelling via homomorphic observations (things which map directly from one example to another), that was then trialled through use, to attempt to falsify it—a scientific method (see Figure 3).

The viable systems model has 5 levels each of which is coupled with its external environment (see Figure 4). The recursion number included in this simplified version allows one to imagine all of the different levels and units, that are themselves viable systems within a greater system of both coupled and encompassing viable systems models.

The 5 sub-systems work as follows (described as pertaining to the firm but could be applied to many different types of systems):

S1: Operations—this is where the work gets done, the sub systems describe work that is being done, that could be done and that the firm would (feasibly) like to do. Bear in mind the recursive model means that each one of these systems is a viable systems model in itself.

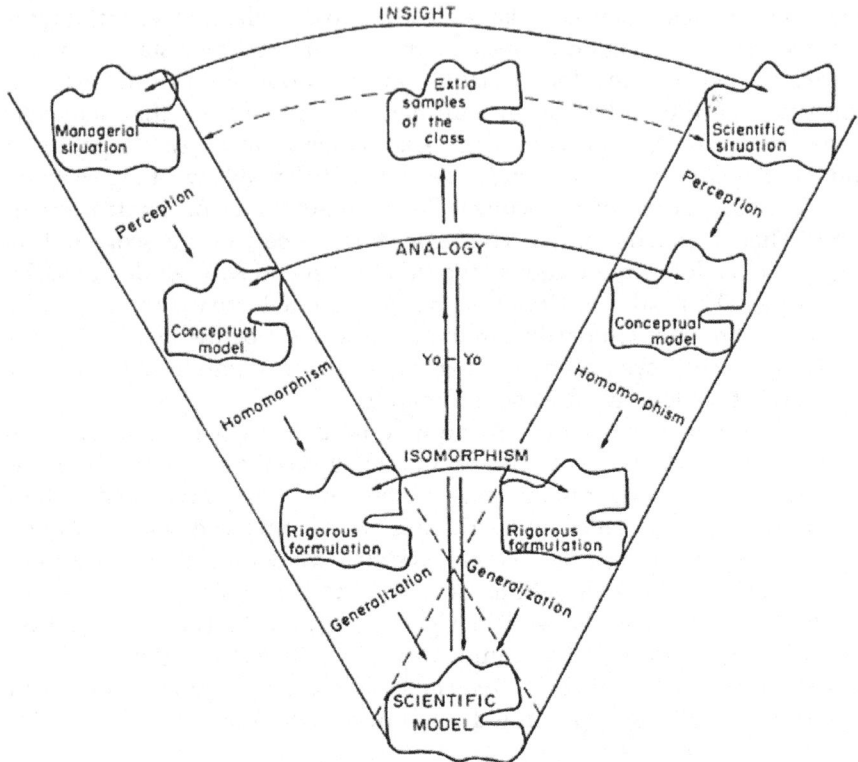

Figure 3. Beer's account of scientific modelling (Beer, 1984).

S2: Co-ordination—co-ordinates system 1, for instance scheduling would occur here, but that scheduling would also be communicated to the wider system via S3. Thus it is all of the activities required to co-ordinate the overall system (at this recursive level).

S3: Optimisation—ensures that the co-ordination of operations is optimal. Systems 1 to 3 are all concerned with the here and now of operational activities.

S4: Development—looks to the future and assesses the impact of externalities on the system.

S5: Valuation—sets the policies, deciding on the normative rules of the organisation. System 5 dictates what must always happen.

Information needs to flow between the systems but it is framed differently due to the different temporal alignments at each level. For instance; system 4 could use benchmarking against external organisations and simulation of possible scenarios to suggest strategy, but system 5 would decide on the best strategy and policy to blow that down through

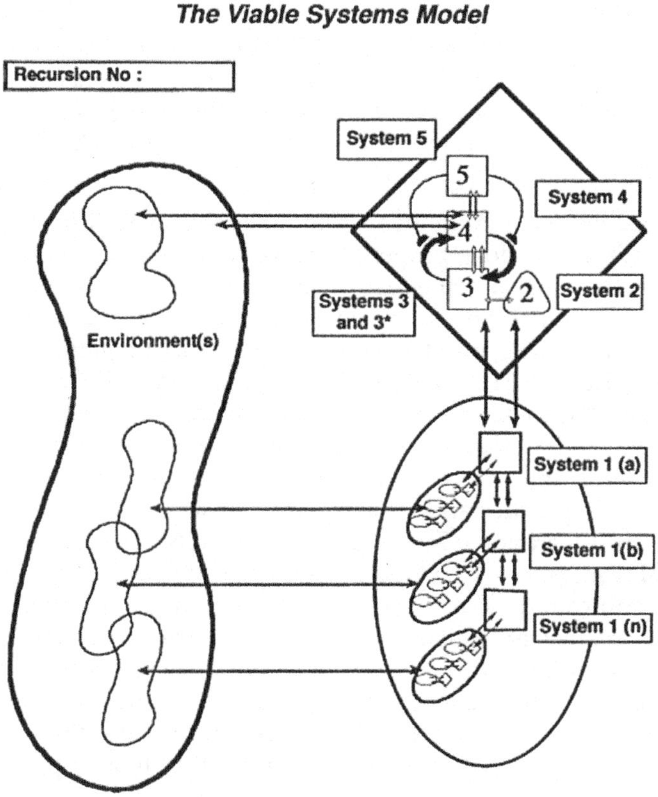

The Viable Systems Model

Figure 4. A simplified VSM, from (Espinosa et al., 2008).

the organisation via communication with system 4, which is in constant communication with 3, while systems 2 and 1 are constantly feeding back to system 3. It looks hierarchical and top down, but only if one forgets the underlying concept of recursion.

There is much more to the Viable Systems Model (VSM) and if the reader is inclined to use this method we would suggest reading Beer (1984) and Espinosa et al. (2008) as a starting point and refer to the Stafford Beer collection at Liverpool John Moores University (https://www.ljmu.ac.uk/microsites/library/special-collections-and-archives/special-collections/stafford-beer-collection).

What we are interested in here is the design thinking aspects of this methodology;

> "What the VSM helps us do is to create more effective organisation by engaging the energy and intelligence of local constituents in the overall endeavour."

"Participation and Re-engagement: From the VSM perspective, participation is so fundamental that it often escapes notice. Variety balancing between all operations at all levels require empowered, engaged individuals/ communities/organisations, and the only way effective organisation can be articulated is to devolve power to the level that things get done" (Espinosa et al., 2008).

Beer did not feel that he needed to make this explicit in the method, if the modelling was done properly this would automatically be the case. The design thinking is embedded in the modelling process as it has to engage across all levels of an organisation in an open and inclusive way and as Brown and Katz put it, design thinking is about *'putting people first'* (Brown and Katz, 2011). To assess viability, all levels must be included, since it is possible for viable systems to exist within and beside non-viable ones. The method is claimed by both systems and by problem structuring methods within the operational research community and continues to be used both in practice and by academics, as was evidenced in a recent survey by the Operational Research Society PSM SIG, presented at OR61, the annual conference of the Operational Research Society (see Figure 5).

The codification of the problem domain into a recognisable and repeatable artefact is similar to the problem structuring methods that we looked at earlier in this chapter. In all of these examples the methods produce artefacts that are not end products or improved designs in themselves but which are repeatable models to aid thinking about problems and domains to enable improvement. In that respect they can

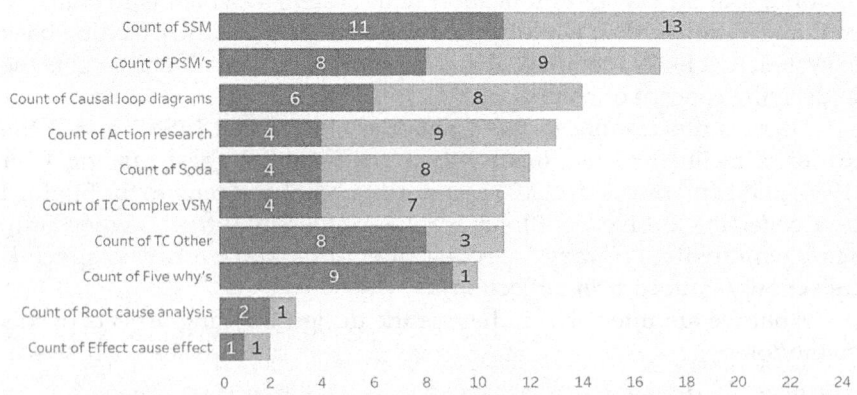

Figure 5. Operational Research Society, Problem Structuring Methods special interest group (presented at OR61 Kent) a review of tools member's use, to deal with complexity. Paler grey is academics, darker is practitioners, count indicates the number of participants answering the question.

both open up and close down creative thinking but that is where the design thinking comes in. The methods all have, often explicitly defined within them, a recognisable design thinking stance, even though it is not called design thinking.

In later work on healthcare systems Beer started to reach out to new ways of modelling systems as he found that the healthcare systems he studied could not be described as viable. He noted that the different parts of the system were responding to different external drivers, as the hospitals sought to mitigate any risk of litigation, the public sought to gain eternal life, and were swayed by quackery (prompted by media and corporate self-interest) and the government sought to reduce expenditure. This prompted him to start trying to define systems in a new way, which looked to define complex system behaviour across both viable and non-viable systems using ideas of flow and iconic labels. These high level descriptors of system behaviour can help us to perform the codification and analysis that enables design thinking, and it could be argued that he was thinking in a highly designerly way in his quest to make complex systems comprehensible and tangible to as wide an audience as possible.

Stafford Beer was multidisciplinary thinker, rooted in practice, an artist, shaman and poet, systems thinker, operational researcher, psychologist and cybernetician. He initially trained in psychology but grew an interest in operational research after his involvement with the army during the Second World War. He discovered cybernetics and started applying it to problems that he was coming across in his management practice which led to his seminal text on management cybernetics and operational research (Beer, 1966). During the 1950's and 60's he became, in many ways, the archetypal management guru whose ideas worked in practice and got results. This all changed after he was invited to help with aspects of the socialist state of Salvador Allende in Chile, by putting in place a cybernetic management system to steer the economy (Medina, 2006). He was deeply affected by the subsequent revolution and the violent deaths and disappearances of many of his former colleagues. He withdrew to a cottage in Wales where he continued to write, teach and perform consultancy until his death in 2002. Beer remained highly influential, and produced many of his bestselling works during this time, but he shunned the high life of his previous existence, preferring a quiet place that nurtured meditation and creativity.

Before we move on from Cybernetics the parallels with the work of Herbert Simon should be mentioned, this passage from his updated book (1997) is particularly interesting; *"In very many cases whether a particular system will achieve a particular goal or adaptation depends on only a few characteristics of the outer environment and not at all on the detail of that environment. Biologists are familiar with this property of adaptive systems under the label of homeostasis. It is an important property of most good designs,* whether biological or artifactual. *In one way or another the designer insulates the inner system from the environment, so that an invariant relation is maintained between inner system and goal, independent of variations over a wide range in most parameters that characterize the outer environment"* (Simon, 1997). What he is saying here is resonant of the cybernetician's black box, for Simon it would be the boundedness of a system. He also talks of homeostasis as a desirable property for a system to possess and even ties this to a measure of *'good design'*. The insulation from the complexity of the environment and consequent *'invariant relation'* to the outer system can be likened to the requisite variety problem that often makes many small simple models/ solutions to parts of a complex and messy problem more desirable than large expansive ones that respond to many external variables (Maier et al., 2014).

Policy research design

This special interest group in the Operational Research Society is the only one that explicitly recognises design (Royston and Komashie, 2016). The group uses many of the problem structuring methods outlined above, including the viable systems modelling approach and the strategic choice approach. They also use techniques, which are more thought of as design thinking tools, such as possible futures and personas. They are staying true to the 'design' in their name by being focussed on what works to allow design thinking that is appropriate to the situation and that is most likely to gain the desired improvement in policy decisions. They are not wedded to any particular method but to what works and as a policy design group that means methods that explore problem domains and create potential solutions in an inclusive, iterative and empathic way.

Another suite of techniques that the group uses regularly is Multi Criterial Decision Methods (MCDM), this is a big subject in operational research with its own groups and many publications. Its interest for us is that it is a mixed methodology that explicitly attempts to bridge the socio-technical and it has been devised to help with decision making in complex and multifaceted environments, the type of messy, ill-structured problem spaces that design and design thinking addresses (Simon, 1997). We now take a closer look at mixed methods and methods that have always had an element of both social and technical.

Mixed methods in Operational Research—bridging the socio-technical

The subject of mixing methods in operational research has a long history as one would expect based on the discussion so far. A few methods explicitly couple these aspects such as system dynamics and multi-criteria decision analysis while other research streams look at using different methods to achieve the upfront scoping of the problem domain so that better modelling can be achieved. The rise of analytics as a subject has bought this mixing of methods to the fore.

Analytics

Mortenson et al. (2015) define analytics as part of a suite of disciplines that make up the paradigm of dianoetic management systems; those based on rational, data driven decision making. They describe the taxonomy that makes up analytics as consisting of technology, quantitative methods, and decision making and behaviour. If we are going to use our analytical models to predict and improve then we also need to be able to achieve robust causal inference. This requires high involvement of subject matter experts, with detailed domain knowledge (Hernán et al., 2019) and techniques to engage and learn with them effectively in bringing technological insights and solutions to bear. This is part of what Burger et al. (2019) describe as smart operational research, an operational research that can address the needs of the socio-technical systems that are becoming more prevalent in the modern workplace.

Modelling

Much of what we call analytics has been studied as modelling, i.e., time series modelling and forecasting, or simulation. Design can be seen as modelling (Maier et al., 2014) and modelling is used as a communication tool in the design process (Eckert et al., 2005). This intertwined view of modelling and design stems from a cybernetic perspective in the field of engineering design (Maier et al., 2014), where the importance of iterations, contextualisation and communication are noted, particularly in systems which couple the social and the technical (Piccolo et al., 2019; Wynn and Eckert, 2017). Models are boundary objects which can potentially align stakeholder perceptions through a negotiated process of cognitive understanding facilitated by the model and this has been noted by both design engineers (Maier et al., 2014), and operations researchers (Carlile, 2004; Franco, 2013; Luoma, 2015). The need for thoughtful design when using modelling as a decision aid is noted by Luoma (2015), who calls for models to attain a *'behavioural fit'* within an organisation by being sympathetic to work pace and structure. He shows how models built in this way are often more successful in implementation due to decreased risk of disruption with associated risks and potential conflict.

Behavioural fit implies the use of stakeholders and methods, which facilitate engagement and structure problem domains, laying bare political and social aspects, as in problem structuring methods and design thinking. Encouraging high engagement can increase the validity of models, encouraging acceptance and even shift cultures (Kotiadis and Mingers, 2006; Pessôa et al., 2015; Phillips and Nikolopoulos, 2019). The appropriate methods to fuse depends upon the circumstances (Groop et al., 2017; Howick and Ackermann, 2011; Pessôa et al., 2015) and practice may not occur in the prescribed manner, which can deter some as they feel they do not have the expertise to try out new methods, or they may be wedded to their 'pet' technique. However, research shows this should not be problematic (Small and Wainwright, 2018) if the methods are allowed and able to flex. This 'choosing what works' and 'trying things out' stance is design thinking in all but name.

Frameworks and guidelines for using multi-methodology approaches have been developed in operational research (Lane and Oliva, 1994; Mingers and Brocklesby, 1997; Pollack, 2009; Tako and Kotiadis, 2015). Some of these articles have used mixed methods to aid information systems redesign or technology implementation (Pollack, 2009; Small and Wainwright, 2018) and a few have created analytical artefacts, mainly in the realm of healthcare (Kotiadis and Mingers, 2006; Kotiadis et al., 2013; Pessôa et al., 2015; Phillips and Nikolopoulos, 2019).

PartiSim

A multi-methodology framework (Tako and Kotiadis, 2015) that couples Soft Systems Methodology (SSM) and Discrete Event Simulation (DES), to create a framework for a highly participative process of simulation model design (a predictive analytics technique). It was developed over the course of multiple interventions that sought to mix methods specifically in a healthcare setting (Kotiadis et al., 2013). The development was prompted by a need to bring together stakeholders and the modelling team so that they might build conceptual models together, which would be more likely to address the underlying issues (Kotiadis and Robinson, 2008). Not only has the method been successfully used to redesign healthcare systems (Tako and Kotiadis, 2015) but it has also been used to help implement business improvement in a complex pharmaceutical provider (Phillips and Nikolopoulos, 2019).

Lane and Oliva (1994) coin the term 'dynamic coherence' to describe the need to maintain consistency between behaviours alternately induced by suggested change and/or suggested causal effects. They were looking at coupling Soft Systems Methodology (SSM) and System Dynamics (SD) utilising the political sensitivity and worldview exploration of SSM to confirm personal and social views, whilst using the modelling of SD to confirm system viability.

System dynamics

This is a modelling system based on stocks and flows, which explicitly couples the soft and the hard sides of operational research. The intial modelling process uses causal maps with the stocks and flows, including directions and flow drivers labelled on them. This process is much like a problem structuring method, although as Lane and Oliva (1994) found, it benefits from having an explicit method to deal with politics and perceptions. The technique is good at handling messy domains where there is feedback and complex interrelations, there have however, been accounts of system dynamics interventions which do not get beyond the causal model as this can often solve the problems it was bought in to explore before any modelling takes place (Mingers and Rosenhead, 2004). Unfortunately these cases are rarely written about in the extant literature as they do not move the theory behind system dynamics forward explicitly. However, one might argue that they do move forward the theory behind the practice and design of system dynamics, and interventions in operational research in general.

The analytical modelling in system dynamics is difficult as it often involves piecewise methods for solution of coupled differential equations. As with strategic options development analysis, system dynamics benefits from using good software and there are many systems that have been developed over the years to make the whole thing more intuitive (i.e., Sysdea; https://app.sysdea.com/models/tutorial). With the advent of modern computing power and memory there is an attempt to make the codification of system dynamics models a more human centric enterprise that lends itself to trial and error, exploration of possibilities, and visually intuitive representations. To make this work an expert facilitator is required who can move the modelling along, but who also listens closely to client feedback and suggestions, thus, it could be argued that, performing this practice using design thinking would embed many of the desired characteristics needed for effective modelling to occur.

Mixed methods

The mixing of otherwise differentiated methods has been used in many operational research case studies, either to create technological end artefacts, or to study and aid attempts at integrating technologies into practice. The multimethod framework of Mingers and Brocklesby (1997), which was developed to be used with any kind of mixing; either within soft operational research practice, i.e., multiple problem structuring methods, or across the paradigms, i.e., soft and hard. This framework has recently been used by Small and Wainwright (2018) to map their own multi-methodology so that it could be re-used, but as they noted; *'The participants of the problem situation are best suited to understand their political and socio-cultural needs'* (Small and Wainwright, 2018). Pessôa et al. (2015) also noted that their intervention may have been more successful if there had been a higher engagement throughout the modelling process.

Another recent Operational Research in a healthcare paper used ethnography alongside Soft Systems Methodology (SSM) and Discrete Event Simulation (DES) noting that it enabled the analyst to become part of an exploratory process using data and reported observations (Lamé et al., 2019), they also felt that long term engagement allowed the analyst to learn the context and to become an accepted part of the process under study, thereby increasing engagement and enhancing communication. Phillips and Nikolopoulos (2019) describe a similar effect when using action research and intervention theory to improve forecasting (considered one of the primary methods in predictive analytics). These observations point toward the need for a broad, overarching, and inclusive socio-technical stance in operational research practice, like design thinking.

MCDM (also called MCDA—Multi Criteria Decision Method/Analysis)

This is a process of eliciting criteria and weightings from participants such that they can be used in an algorithmic decision making process that helps to define the best choice from a set of alternatives. It requires one to;

- *'Integrate objective measurement with value judgement;*
- *Make explicit and manage subjectivity'* (Belton and Stewart, 2002).

There are many methods to perform the analytic calculations once the needs have been elicited and there are also many methods and papers written about the needs elicitation process. We would recommend the Belton and Stewart book (Belton and Stewart, 2002) for any interested readers. The application of multi-criteria decision analysis methods are inherently socio-technical processes and have many parallels with design thinking if performed in tandem with a design in practice (the specific multi-criteria decision method, e.g., TOPSIS, AHP among others).

Multi paradigm mixed methods, as in socio-technical systems, are both desirable and necessary according to Mingers and Brocklesby (1997). In the realm of analytics, mixing positivist and interpretive paradigms (e.g., hard and soft) is unavoidable as human users come together with increasingly sophisticated computer provided knowledge (Thomson et al., 2014). Design thinking is a design paradigm that uses a framing approach to prototype designs, which can then be iterated upon to achieve a better end (Dorst, 2011), it is implicit that this will be done in a participative way that co-evolves solutions to messy problematic domains and that many of these problems will include analytical/technical artefacts (Dorst, 2019). Designers are not so concerned with the paradigm in which they are operating at a philosophical level as the only paradigm of concern is one of design. Perhaps operational research would find it easier to pursue mixed methods and multiparadigm practice if they looked at it from a design thinking point of view, and perhaps design as a discipline could use some of the techniques and learning from operational research mixed methods practice?

Conclusion

To wrap up this chapter on operational research and design thinking we revisit the question posed in the introduction; how do the ways of knowing, as outlined in Cross (1982), manifest in operational research practice, explicitly or otherwise?

o Designers tackle 'ill-defined' problems

For almost all of the methods we have covered in this chapter, particularly those defined as soft operational research and problem structuring methods, a primary driver has been the ability to tackle ill-defined and messy problems. The problem structuring methods attempt to tackle these with various techniques to elicit discourse around personal constructions (cognitive mapping) and strategic goals (SODA, JOURNEY making and Strategic Choice Approach). In the case of Soft Systems Methodology (SSM) the construction of rich pictures and root definitions around the CATWOE process is specifically designed to structure a messy domain in a participative and evolutionary way. One of the most used parts of SSM is the CATWOE and in particular the opening up of individual worldviews (the W of CATWOE).

The techniques of Management Cybernetics reduce the messy problem domain to what Simon calls bounded problems, defined by input and output parameters that can be treated as black boxes. This way of viewing an ill-defined space addresses the requisite variety problem and enhances the modelling of domains such that they become amenable to both understanding and structuring. The use of Viable Systems Modelling in a recursive and iterative fashion allows domains to be broken down into parts, each of which can be studied and handled on its own merits.

All of the methods that we looked at in this chapter are for use in either purely social or sociotechnical domains. In this regard, they are always looking to refine ill-defined spaces into more structured pictures from which the right questions to ask, and methods of solution can follow.

- Their mode of problem-solving is 'solution-focused'

Problem structuring methods have broken down into different areas such that they might specifically provide solutions for strategy, for group negotiation, or to structure problematic domains. In this regard they are looking to provide solutions for specific problem types. Soft operational research is broader in its definition in that it is any method that helps us deal with the individual and social issues around problems and messy domains, the idea being that it should help all involved actors to work toward solutions. It has been noted in some simulation work that the upfront soft part of the exercise can sometimes provide a solution such that the simulation never needs to be built. The operational researcher does not then insist the model should be made, rather they are satisfied that the client has come to a solution. In this respect operational researchers are just like designers.

- Their mode of thinking is 'constructive'

Many of the methods we have looked at begin to move away from design thinking as they attempt to provide a highly structured constructive output that can be used for negotiation (as in SODA or JOURNEY making), or for strategy options (as in Strategic Choice Approach). Multi-criteria decision methods use problem structuring and soft approaches to elicit both criteria for decisions and weightings of importance for those criteria, they are constructing the data for input into specific methods. The very heart of operations research as *'the science of better'* requires constructive thinking which is always working toward improvement, and often an elicitation of parameters for very specific methods of solution are required. This can cause a tension between the needs of the model and the views of the users but most operational researchers will try to find creative ways to satisfy the normative structure of algorithmic constraints. This is a very designerly way of behaving; both creative and normative, while trying to move stakeholders toward a defined construction that will provide a solution.

- They use 'codes' that translate abstract requirements into concrete objects

All methods of operational research, which structure domains (such as Soft Systems Methodology), individual thinking (such as cognitive maps), or which elicit requirements do this. As we have to work toward a normative structure we still want to harness the creativity and world views of individual stakeholders. Providing well-structured codes for each methodology enables all of us as a group to move from the abstract toward the concrete. Although we have taken a very rapid tour through operational research it is clear that each methodology has its own 'codes', for instance Viable Systems Model, has an associated language about systems and levels, and it has associated procedures such as recursive application and iterations. All of these sit alongside a highly defined visual model and the data required to feed it.

Even those methods that are soft yet not as defined as problem structuring (such as action research or systems thinking) have defined codes. These are methods that are not as structured making them ideal for exploratory work, where the possible solutions are still far from sight, or where the questions that need to be asked have not been well formed. They foster a design thinking approach in that they need moments for reflection and require high human involvement. The 'codes' of these methods are high level providing broad bounded and less normative approaches.

- They use these codes to both 'read' and 'write' in 'object languages'

Soft operational research provides a nice instance to explore here as the connection is less obvious. Research oriented action research, may not have highly defined constructs but it does have some well-defined constitutive rules (codes, object language), such as iterations of development, the need for a group decision about an action, the observation of that action, and the time to reflect, as a group, upon the implementation of that action. To become research oriented action research, a single defined epistemic lens (object language) is required through which to view multiple instances of the same type of intervention. This allows comparison across cases such that theory can emerge, in this way the object language becomes part of a research design tool.

Operational research then, is a designerly activity since it often follows all of the designerly ways as laid out by Cross (1982). Although not all of operational research uses design thinking explicitly, even the harder mathematical methods are strongly coded, solution oriented and have their own object language, thus they are designerly in their way of thinking. The harder operational research methods are at the interface of scientific ways of thinking and designerly ways, as much of engineering is. The soft operational research methods, problem structuring approaches and mixed methods approaches often create knowledge at the nexus of practice, involve stakeholders throughout the process, are reflective and iterative, which is a design thinking mode of practice. Many of these methods also use designerly ways due to having strong constructs, defined codes and their own object language.

The future of design in operational research

With the advent of ubiquitous high powered computing and readily available, data heavy analytics techniques, the coupling of social systems with these domains has become a prescient subject. Often we need to use operational research and analytics to help solve wicked problems, or at least to give insight into possible solutions in these domains. Wicked problems is part of what design thinking is good at tackling (Buchanan, 1992) and operational research is evolving a smart operational research(Burger et al., 2019) through the use of hybrid practice theories.

In a forthcoming paper, Phillips (2021), describes what she calls *'Human Centric Analytics (HCA), a design paradigm for the human centred design of analytics'*, and she notes the use of design thinking within the framework developed for implementation. Previously the nearest operational research has come to human centric design of analytics is the Business Analytics Methodology (BAM) (Hindle and Vidgen, 2017). The authors' see BAM as an opportunity to re-evaluate a business model and point to a co-evolution of analytical applications via the interaction between data scientist and the business analyst. Co-evolution is a facet of design thinking (Dorst, 2019) and has been noted as a requirement for the development of analytics that give value in organisations (Vidgen et al., 2017).

It would seem from this review of operational research methods and recent developments that the time has come to bring together operational research and design thinking in a more explicit and deliberate manner. It would also be good to see more research on the links between operational research and design, and a deeper investigation into the use of *'designerly ways'* (Cross, 1982) in operational research practice.

References

Ackermann, Fran and Colin Eden. 1989. Strategic options development and analysis. pp. 135–90. *In*: Holwell, Sue and Reynolds, Martin [eds.]. Systems Approaches to Managing Change: A Practical Guide. John Wiley and Sons Limited. Chichester, UK.

Ackermann, Fran. 2012. Problem structuring methods 'in the Dock': Arguing the case for soft OR. European Journal of Operational Research, 219(3): 652–58.

Ackoff, Russell L. 1979. The future of operational research is past. Journal of Operational Research Society, 30(2): 93–104.

Ackoff, Russell L., Jason Magidson and Herbert J. Addison. 2006. Idealized Design: How to Dissolve Tomorrow's Crisis Today. Prentice Hall, New Jersey, USA.

Argyris, Chris. 1970. Intervention Theory & Method: A Behavioral Science View. Addison-Wesley Publishing Company, Reading, Masaacheusettes, USA.

Ashby, William Ross. 1956. An Introduction to Cybernetics. London, 1956. Internet (1999): Chapman & Hall, London, UK.

Baskerville, Richard L. and Trevor Wood-Harper, A. 1996. A critical perspective on action research as a method for information systems research. Journal of Information Technology, 11(3): 235–46.
Baskerville, Richard and Richard Baskerville. 2008. What design science is not. European Journal of Information Systems, 17(5): 441–443.
Beer, Stafford. 1966. Decision and Control: The Meaning of Operational Research and Management Cybernetics. 1st ed. John Wiley & Sons Ltd., Chichester, UK.
Beer, Stafford. 1984. The viable system model: Its provenance, development, methodology and pathology. Journal of the Operational Research Society, 35(1): 7–25.
Belton, Valerie and Theodor J. Stewart. 2002. Multiple Criteria Decision Analysis: An Integrated Approach. Kluwer Academic Publishers.
Brown, Tim and Barry Katz. 2011. Change by design. Journal of Product Innovation Management, 28(3): 381–83.
Buchanan, Richard. 1992. Wicked problems in design thinking. Design Issues, 8(2): 5.
Buchanan, Richard. 2019. Systems thinking and design thinking: The search for principles in the World we are making. She Ji, 5(2): 85–104.
Burger, Katharina, Leroy White and Mike Yearworth. 2019. Developing a smart operational research with hybrid practice theories. European Journal of Operational Research, 277(3): 1137–50.
Caniato, F., Kalchschmidt, M. and Ronchi, S. 2011. Integrating quantitative and qualitative forecasting approaches: organizational learning in an action research case. Journal of the Operational Research Society, 62(3): 413–24.
Carlile, Paul R. 2004. Transfering, translating, and transforming: An integrative framework for managing knowledge across boundaries. Organization Science, 15(5): 555–568.
Checkland, Peter. 1985. From optimizing to learning: A development of systems thinking for the 1990s. Journal of the Operational Research Society, 36(9): 757–67.
Checkland, Peter and Holwell, S. 1998. Action research: Its nature and validity. Systemic Practice and Action Research, 11(1): 9–21.
Checkland, Peter. 1999. Systems Thinking, Systems Practice: Includes a 30-Year Retrospective. Wiley, Chichester, UK.
Checkland, Peter. 2000. Soft systems methodology: A thirty year retrospective. Systems Research and Behavioral Science, 17: 11–58.
Churchman, C. West, Russell L. Ackoff and Leonard E. Arnoff. 1957. Introduction to Operations Research. Wiley, New York, USA.
Churchman, C. West. 1971. The Design of Inquiring Systems: Basic Concepts of Systems and Organization. Basic Books, New York, USA.
Churchman, C. West. 1979. The Systems Approach and Its Enemies. Basic Books, New York, USA.
Ciccone, N., Patou F., Komashie A., Lamé G., Clarkson P.J. and Maier A. 2020. Healthcare systems design: a sandbox of current research themes presented at an international meeting. In Proceedings of the Design Society (Vol. 1, pp. 1873–1882). [163] Cambridge University Press. The DESIGN Conference https://doi.org/10.1017/dsd.2020.24.
Cross, Nigel. 1982. Design as a discipline: Designerly ways of knowing. Design Studies, 3(4): 221–27.
Cross, N. 2007. From a design science to a design discipline: understanding designerly ways of knowing and thinking. *In*: Michel, R. [eds.]. Design Research Now. Board of International Research in Design. Birkhäuser Basel. https://doi.org/10.1007/978-3-7643-8472-2_.
Dorst, Kees. 2011. The core of 'Design Thinking' and its application. Design Studies, 32(6): 521–32.
Dorst, Kees. 2019. Co-evolution and emergence in design. Design Studies, 65: 60–77.
Eckert, Claudia, Anja M. Maier and Chris McMahon. 2005. Communication in design. Design process improvement: A Review of Current Practice, pp. 232–61.

Eden, Colin. 1988. Cognitive mapping. European Journal of Operational Research, 36(1): 1–13.
Eden, Colin and Chris Huxham. 2006. Researching organizations using action research. pp. 388–408. *In*: Clegg, S.R., Hardy, C., Lawrence, T.B. and Nord, W.R. [eds.]. The Sage Handbook of Organization Studies. Sage, London, UK.
Eden, Colin and Fran Ackermann. 2018. Theory into practice, practice to theory: Action research in method development. European Journal of Operational Research, 0: 1–11.
Espinosa, A., Harnden, R. and Walker, J. 2008. A complexity approach to sustainability—stafford beer revisited. European Journal of Operational Research, 187(2): 636–51.
Fischer, Thomas and Christiane M. Herr [eds.]. 2019. Design Cybernetics. Springer International Publishing, Cham.
Franco, L. Alberto. 2013. Rethinking soft or interventions: Models as boundary objects. European Journal of Operational Research, 231(3): 720–33.
Friend, J. and Jessop, N. 1976. Local Government and Strategic Choice, 2nd Edition. Pergamon Press, Oxford, UK.
Friend, J. 1992. New directions in software for strategic choice. European Journal of Operational Research, 61(1-2): 154–64.
Friend, J. and Hickling, A. 2012. Planning under Pressure: The Strategic Choice Approach, Third Edition. 3rd ed. Pergamon Press, Oxford, UK.
Groop, Johan, Mikko Ketokivi, Mahesh Gupta and Jan Holmström. 2017. Improving home care: Knowledge creation through engagement and design. Journal of Operations Management, 53–56(November): 9–22.
Hernán, Miguel A., John Hsu and Brian Healy. 2019. A second chance to get causal inference right: A classification of data science tasks. Chance, 32(1): 42–49.
Hevner, Alan R., Salvatore T. March, Jinsoo Park and Sudha Ram. 2004. Design science in information systems research. MIS Quarterly, 28(1): 75–105.
Hindle, Giles A. and Richard Vidgen. 2017. Developing a business analytics methodology: A case study in the Foodbank sector. European Journal of Operational Research, 0: 1–16.
Howick, Susan and Fran Ackermann. 2011. Mixing or methods in practice: past, present and future directions. European Journal of Operational Research, 215(3): 503–11.
Jackson, M.C. and Keys, P. 1984. Towards a system of systems methodologies. The Journal of the Operational Research Society, 35(6): 473–86.
Jackson, M.C. 1993. Social theory and operational research practice. The Journal of the Operational Research Society, 44(6): 563–77.
Jackson, M.C. 2001. Critical systems thinking and practice. European Journal of Operational Research, 128(2): 233–44.
Jackson, M.C. 2009. Fifty years of systems thinking for management. Journal of the Operational Research Society, 60(SUPPL. 1).
Keys, Paul. 2000. Creativity, design and style in MS/OR. Omega, 28(3): 303–12.
Kimbell, Lucy. 2011. Rethinking design thinking: Part I. Design and Culture, 3(3): 285–306.
Kimbell, Lucy. 2012. Rethinking design thinking: Part II. Design and Culture, 4(2): 129–48.
Kirby, M.W. 2003. The intellectual journey of Russell Ackoff: From OR Apostle to OR Apostate. Journal of the Operational Research Society, 54(11): 1127–40.
Kotiadis, Kathy and John Mingers. 2006. Combining PSMs with hard or methods: The philosophical and practical challenges. Journal of the Operational Research Society, 57(7): 856–67.
Kotiadis, Kathy and Stewart Robinson. 2008. Conceptual modelling: Knowledge acquisition and model abstraction. pp. 951–58. *In*: Mason, S.J., Hill, R.R., Mönch, L., Rose, O., Jefferson, T. et al. [eds.]. Proceedings of the 2008 Winter Simulation Conference.
Kotiadis, Kathy, Antuela A. Tako and Vasilakis, C. 2013. A participative and facilitative conceptual modelling framework for discrete event simulation studies in healthcare. Journal of the Operational Research Society, 65(2): 197–213.
Krippendorff, Klaus. 2007. The cybernetics of design and the design of cybernetics. Kybernetes, 36(9–10): 1381–92.

Kuhn, Thomas S. 1962. The Structure of Scientific Revolutions.
Lamé, Guillaume, Oualid Jouini and Julie Stal-Le Cardinal. 2020. Combining soft systems methodology, ethnographic observation, and discrete-event simulation: A case study in cancer care. Journal of the Operational Research Society, 71(10): 1545–1562.
Lane, David C. and Rogelio Oliva. 1994. The greater whole: Towards a synthesis of SD and SSM. 1994 International System Dynamics Conference. Problem—Solving Methodologies, pp. 134–145
Luoma, Jukka. 2016. Model-based organizational decision making: A behavioral lens. European Journal of Operational Research, 249(3): 816–826.
Maier, A.M., Wynn D.C., Andreasen M.M. and Clarkson P.J. 2012. A cybernetic perspective on methods and process models in collaborative designing. pp. 233–240. In: Marjanovic Dorian, Storga Mario, Pavkovic Neven and Bojcetic Nenad [eds.]. DS 70: Proceedings of DESIGN 2012, the 12th International Design Conference, Dubrovnik, Croatia.
Maier, Anja M., David C. Wynn, Thomas J. Howard and Mogens Myrup Andreasen. 2014. Perceiving design as modelling: A cybernetic systems perspective. pp. 133–49. In: Chakrabarti, A. and Blessing, L. [eds.]. An Anthology of Theories and Models of Design: Philosophy, Approaches and Emprical Explorations. Springer.
McDonald, A.G., Cuddeford, G.C. and Beale, E.M.L. 1974. Balance of care: Some mathematical models of the national health service. British Medical Bulletin, 30(3): 262–70.
Medina, Eden. 2006. Designing freedom, regulating a nation: Socialist cybernetics in Allende's Chile. Journal of Latin American Studies, 38(3): 571–606.
Midgley, Gerald, Robert Y. Cavana, John Brocklesby, Jeff L. Foote, David R.R. Wood et al. 2013. Towards a new framework for evaluating systemic problem structuring methods. European Journal of Operational Research, 229(1): 143–54.
Mingers, John and Sarah Taylor. 1992. The use of soft systems methodology in practice. Journal of the Operational Research Society, 43(4): 321–32.
Mingers, John and John Brocklesby. 1997. Multimethodology: Towards a framework for mixing methodologies. Omega, 25(5): 489–509.
Mingers, John and Jonathan Rosenhead. 2004. Problem structuring methods in action. European Journal of Operational Research, 152(3): 530–54.
Mingers, John. 2011. Soft OR comes of age-but not everywhere! Omega, 39(6): 729–41.
Mortenson, Michael J., Neil F. Doherty and Stewart Robinson. 2015. Operational research from taylorism to terabytes: A research agenda for the analytics age. European Journal of Operational Research, 241(3): 583–95.
Mumford, Enid. 2006. The story of socio-technical design: Reflections on its successes, failures and potential. Information Systems Journal, 16(4): 317–42.
O'Keefe, Robert M. 1995. MS/OR enabled systems design. Operations Research, 43(2): 199–207.
O'Keefe, Robert M. 2014. Design science, the design of systems and operational research: back to the future ? Journal of the Operational Research Society, 65: 673–84.
Pask, Gordon. 1969. The architectural relevance of cyberspace. Architectural Design, (7/6): 68–77.
Pessôa, Leonardo Antonio Monteiro, Marcos Pereira Estellita Lins, Angela Cristina Moreira Da Silva and Roberto Fiszman. 2015. Integrating soft and hard operational research to improve surgical centre management at a University Hospital. European Journal of Operational Research, 245(3): 851–61.
Phillips, Christina Jane. 2021. When simulation becomes human centric analytics. pp. 164–167. In: Fakhimi, M., Robertson, D. and Boness, T. [eds.].Proceedings of the Operational Research Society Simulation Workshop 2021 (SW21). Operational Research Society.
Phillips, Christina Jane and Konstantinos Nikolopoulos. 2019. Forecast quality improvement with action research: A success story at PharmaCo. International Journal of Forecasting, 35(1): 129–43.

Piccolo, Sebastiano A., Anja M. Maier, Sune Lehmann and Chris A. McMahon. 2019. Iterations as the result of social and technical factors: empirical evidence from a large-scale design project. Research in Engineering Design, 30(2): 251–70.

Pickering, Andrew. 2002. Cybernetics and the mangle. Social Studies of Science, 32(June 2002): 413–37.

Pickering, Andrew. 2009. Beyond design: Cybernetics, biological computers and hylozoism. Synthese, 168(3): 469–91.

Pollack, J. 2009. Multimethodology in series and parallel: Strategic planning using hard and soft Or. Journal of Operational Research Society, 60: 156–67.

Ranyard, J.C., Fildes, R. and Tun-I. Hu. 2015. Reassessing the scope of OR practice: The influences of problem structuring methods and the analytics movement. European Journal of Operational Research, 245(1): 1–13.

Royston, G. 2013. Operational research for the real world: Big questions from a small Island. Journal of the Operational Research Society, 64(6): 793–804.

Royston, G. and Komashie, A. 2016. Full STEAM Ahead for Operational Research and Design? P. 82 in OR58 Annual conference, Portsmouth.

Royston, G., Halsall, J., Halsall, D. and Braithwaite, C. 2003. Operational research for informed innovation: NHS direct as a case study in the design. Implementation and Evaluation of a New Public Serice, 54(10): 1022–28.

Senge, Peter M. and John D. Sterman. 1992. Systems thinking and organizational learning: Acting locally and thinking globally in the organization of the future. European Journal of Operational Research, 59(1): 137–50.

Simon, Herbert A. 1997. The Sciences of the Artificial. 3rd ed. MIT Press.

Small, Adrian and David Wainwright. 2018. Privacy and security of electronic patient records—tailoring Multimethodology to Explore the Socio-Political problems associated with role based access control systems. European Journal of Operational Research, 265(1): 344–60.

Sodhi, M.S. and Christopher S. Tang. 2008. The OR/MS Ecosystem: Strengths, Weaknesses, Opportunities, and Threats. Operations Research, 56(2): 267–77.

Tako, Antuela A. and Kathy Kotiadis. 2015. PartiSim: A multi-methodology framework to support facilitated simulation modelling in healthcare. European Journal of Operational Research, 244(2): 555–64.

Thomson, Robert, Christian Lebiere and Stefano Bennati. 2014. Human, model and machine. pp. 27–31. In: Proceedings of the 2014 Workshop on Human Centered Big Data Research. Raleigh, NC, USA.

van Aken, Joan, Aravind Chandrasekaran and Joop Halman. 2016. Conducting and publishing design science research: inaugural essay of the design science department of the journal of operations management. Journal of Operations Management, 47-48: 1–8.

Vidgen, Richard, Sarah Shaw and David B. Grant. 2017. Management challenges in creating value from business analytics. European Journal of Operational Research, 261(2): 626–39.

Wynn, David C. and Claudia M. Eckert. 2017. Perspectives on iteration in design and development. Research in Engineering Design, 28(2): 153–84.

CHAPTER-9

Design Thinking and Welfare Technology
A Focus on Information Design

Riika Saurio, Lea Hennala, Satu Pekkarinen and Helinä Melkas*

Introduction

Since early definitions of information design (Easterby and Zwaga, 1984; Tufte, 1990), the field has seen a rise in importance and become more multi-faceted in line with, for example, increasing digitalization (Black et al., 2017). Digital transformation has affected everyday life in numerous ways, transforming the ways in which people interact, communicate, work and study. In care services, in line with this transformation, the use of welfare technology is increasing rapidly. Implementation of such technology requires careful planning, including skilled information design for information to be given to and collected from different kinds of users.

Various schools of thought exist in information design research, but there appears to be a consensus about its strong interdisciplinary character. According to Pettersson (1998), good information design "makes everyday life easier for people, and it grants good credibility to the senders or sources". Information design aims at transforming data into high-quality information, and drawing from this, Simlinger (2014) formulated a motto: "High-quality information empowers people to attain goals". In this study, we report the results of research on welfare technology and conclude with

---

Lappeenranta-Lahti University of Technology LUT, School of Engineering Science, Mukkulankatu 19, 15210 Lahti, Finland.
Emails: riika.saurio@lut.fi, lea.hennala@lut.fi, satu.pekkarinen@lut.fi
* Corresponding author: helina.melkas@lut.fi

needs for information design for the successful implementation and use of such technology.

The information design perspective in this chapter is essentially part of design thinking as defined by Roberts et al. (2016) in the context of healthcare management and innovation: "Design thinking is, at its core, a systematic innovation process that prioritizes deep empathy for end-user desires, needs and challenges to fully understand a problem in hopes of developing more comprehensive and effective solutions". While this chapter does not focus on all elements of information design (notably, aesthetic and artistic aspects are beyond its scope), it is concerned with "the defining, planning, and shaping of the contents of a message and the environments it is presented in with the intention of achieving particular objectives in relation to the needs of users" (Pettersson, 1998). This chapter is concerned with empowering people—notably care professionals—to attain goals with the help of information design. The focus is not on the profession(s) of design but on the roles of ordinary working people such as professional caregivers and care managers in information design in workplaces. The focus is also on end-users.

As noted by Taneva (2019), the kind of information supplied can significantly influence the actions taken and, thus, the outcome of interaction. Taneva (2019) asks what the optimal mode of information transmission is between a sender and a group of interacting agents (receivers) who form their beliefs and take actions based on the information provided. While an optimal mode is not sought, in this qualitative analysis, the aim is to bring up certain elements and attributes for appropriate information transmission to implement and orient towards welfare technology use in care services. In contrast to earlier studies, professional caregivers and care managers are perceived as both senders and receivers who operate in a double role. The practically informed starting point in this chapter is that in today's working life, information design is everybody's business. Rather than one designer as a sender, communicating with multiple receivers in interactions, it is often a question of dealing with increasingly varied information design contexts. As compared to many earlier studies on information design, the limits may thus be stretched somewhat.

This is, however, appropriate, as information design is dealt with in a relatively novel area: the use of welfare technology. This concept has emerged to describe a broad field of technology that has earlier been referred to as assistive technology or gerontechnology, for example. It has been defined as follows: "Welfare technology is all technology which in one way or another improves the lives of those who need it. The technology is used to maintain or increase security, activity, participation or independence for people with a disability or the elderly" (Nordic Welfare Centre, 2020). Importantly, this technology also serves care professionals and may be used by older or disabled people's relatives in informal care. The definition of welfare technology has evolved from focussing on applications to including the application, system and administration of service; however, this definition is still under discussion (Östlund et al., 2015).

This chapter contributes to information design research by proactively unveiling needs for information design, in line with Stahl-Timmins' (2017) call for research results in information design (in contrast to commercial information design projects) that may lead to advancement in the field. Attributes of high-quality information are utilized in this study as the framework of analysis (Simlinger, 2014; Wang and Strong, 1996) and the tool for examining information design issues in the context of services in which welfare technology is to be used. The study is based on empirical data collected in Finnish social and healthcare services.

The data collection followed basic principles of design thinking. End-users and their family members and carers were involved, and their voices were heard in the development process (Pfannstiel and Rasche, 2019; Gammon et al., 2014). Although there were certain characteristics of a process improvement orientation, the focus was on the design thinking orientation (see Table 1). It was also in line with the human-centredness of (health) design thinking (Ahookire et al., 2020). They further emphasize the need for a creative mindset as another fundamental principle in health design thinking. This was realized in the present research by producing—with the help of information visualization—discussion cards and an animation on welfare technology utilization in social and healthcare services for use in workplaces and education to advance smoother implementation and use of welfare technology. Design thinking was thus not applied like designers would apply it, leading ideally to an optimized solution to the situation in question, ready for scaled implementation

Table 1. Design thinking versus process improvement approaches (source: Roberts et al., 2016).

Process improvement	Design thinking
Prioritizes evaluation of a limited set of possible solutions	Prioritizes comprehensive understanding of underlying problems
Well suited to address problems that have predictable solutions	Well suited to address problems that have unpredictable solutions (wicked problems)
Promotes consensus building (convergent)	Promotes opposing ideas and debate (divergent)
Aims to uncover what is important to consumers within a particular experience	Aims to uncover what is important to consumers in their everyday lives
Empathy research focuses on what people think to reveal improved outcomes	Empathy research focuses on what people feel to reveal new/disruptive outcomes

(Roberts et al., 2016; Altman et al., 2018), but according to what was actionable for information design researchers.

Information design related to welfare-technology-related services is a challenging task that requires—apart from design thinking—a good informational basis in increasingly complex operational environments. Given the increasing importance of successful embedding of technological innovations, this research is likely relevant for those interested broadly in information, design, technology, and management.

Design thinking and information design for welfare technology

The goal in information design should always be clarity of communication; the message must be accurately developed and transmitted by the sender, and then correctly interpreted and understood by the receiver (Pettersson, 1998; Jacobson, 1999). This is simple yet challenging, even in very clearly defined communication. Information is of high quality if it is fit for its intended use in conducting business operations, decision making and planning (Pierce et al., 2006; Melkas et al., 2010). In a policy paper, Simlinger (2014) noted that when speaking about 'high-quality information', the attributes of high quality should be kept in mind (Wang and Strong, 1996). The paper listed the following adjectives: accessible, appropriate, attractive, believable, complete, concise, errorless, interpretable, relevant, objective, timely, secure, understandable and valuable (Simlinger, 2014).

Lee et al. (2002) provided an extensive review of academics and practitioners' views of information quality attributes. In early studies in the field, attributes were developed empirically from information consumers, such as in the Wang and Strong study (1996). Another type developed their attributes from literature reviews, e.g., Goodhue (1995). By grouping all measures from other authors together, they hoped to cover all aspects of the information quality construct. The third type of study focussed on a few dimensions that can be objectively defined (e.g., Wand and Wang, 1996). Practitioners, again, reported the dimensions and measures they use within organizations. The approach is generally not rigorous from a research viewpoint, but it provides some insight into the views of practitioners, such as specialists within organizations, outside consultants and vendors of information products. As they focus on specific organizational problems, coverage of all attributes of high-quality information is not their primary intent. The attributes employed are driven by the context in which they are operating. The contexts have typically included data warehouse development, information quality tools for improving the quality of data input to databases and environments with multiple incompatible databases (Melkas et al., 2010). There is, however, a wealth of other contexts in which this approach is viable, including care services.

In this study, the particularly comprehensive framework of Wang and Strong (1996) is utilized to identify and disaggregate needs for information design. Dyson (2017) presented—as different types of information design studies—the use of frameworks (e.g., information design studies that focus on the organization and structure of information in websites or the use of visuals in ergonomic journal articles); heuristics (e.g., evaluation of senior-friendly websites, expert opinions); diagnostic testing (e.g., testing graphs of social mobility); user research (e.g., evaluation of medicine package leaflets); and other research studies. The present study may be characterized as being located on the interface of the use of frameworks on the one hand and expert opinions as well as user research on the other. As compared to the examples given by Dyson (2017), however, this study has a broader approach along the lines of design thinking in healthcare (e.g., Altman et al., 2018).

Altman et al. (2018) noted that design thinking is similar to both user-centred design and human-centred design. In this study, it provides the overall approach for the study of information design. It was also applied especially in data collection by involving welfare technology users with different roles (Pfannstiel and Rasche, 2019; Gammon et al., 2014), as well as in the later stages in which the tangible outcomes—discussion cards and an animation—were produced for use by practitioners and educators. Among the numerous different characterizations of design thinking, Chasanidou et al. (2015) regarded it as a system of three overlapping spaces in which viability refers to the business perspective of design thinking, desirability reflects the user's perspective and feasibility encompasses the technology perspective. According to them, innovation increases when all three perspectives are addressed. What needs to be added, however, is that these spaces affect each other and change over time; desirability and feasibility, for instance, may change depending on how well technology-related information design succeeds.

According to Horn (1999), it is not more information that is needed but the ability to present the right information to the right people at the right time in the most effective and efficient form. Professionalization of information design is also needed due to the increasing costs of time, including management, technical and professional. Passini (1999) noted that information design "emphasizes communication and is as concerned with content as with form. It has its roots in a variety of disciplines—including information theory and the cognitive sciences—and brings together design and research".

Welfare technology in care services

In earlier research, it has been found that the use of technological devices requires development of service systems as well as care staff's abilities, knowledge and work practices (Melkas et al., 2020b). In addition, it may require a lot of flexibility, adaptation and learning from clients (referring in this study to older people who are clients of social and healthcare services). Still, research and development of technology is often too focussed on devices, and prerequisites of successful, effective application are not investigated in a comprehensive manner. Studying technology as a separate 'island' does not, however, respond to the questions that should be asked at workplaces when introducing new technology, pondering upon work practices later on or when problems arise in, for instance, coping at work. Nor does it respond to the questions that older people and their close ones may have when they search for appropriate welfare technology for the older user and acquire, use and maintain it. Poorly learned practices and inabilities are destructive with regard to well-being at home or in the workplace (Melkas, 2013).

In addition to clients, caregivers in elderly care services may be relatively aged. Their readiness to use technology may be low, especially since welfare technology issues have not been incorporated into their studies. The feeling of not wanting or having enough strength to learn new things may even lead to premature retirement. Correspondingly, the readiness of older people to use technology may be low. Older people and their close ones rarely have information about welfare technology and ways to acquire and use it. It is difficult to assess whether a certain technology is suitable for one's own use (Melkas, 2013). On the other hand, welfare technology users, such as older people, are not a homogeneous group with a static identity (Eriksson, 2016; Östlund et al., 2015). This also concerns caregivers—recent studies concerning care robots, for example, have shown that attitudes do not necessarily depend on age (Turja, 2019). At any rate, welfare technology should not be treated as a separate entity in any environment (Melkas, 2013).

A number of welfare technologies have been used in elderly care services for a relatively long time already (Bouma, 1998), albeit using different concepts. In recent years, more and more technological devices, including more and more robotics, have been introduced to the market. The wide sphere of welfare technology exacerbates challenges for information design due to very different characteristics, uses and users. Research on information design—with a design thinking approach—is thus topical in order to gain the benefits from welfare technology use and to avoid pitfalls.

Implementation of welfare technology

According to Linton (2002), implementation encompasses all functions that take place between commitment to implement a product on the one hand and skilled and consistent use of the product on the other. Implementation is thus a broad process; it needs to be planned and organized well to be successful (Johansson-Pajala et al., 2019). Implementation may take different forms depending on the context (Holm, 2013). It starts from planning and communication; good planning and communication well in advance prepare the work community for the change. The importance of providing feedback and addressing problems should be emphasized. The importance of the organization becomes pronounced in the implementation because the organization decides on the system to be taken into use and is responsible for resources and support for personnel. The most important factors from an individual perspective are the person's attitude and interest, as well as participation in trainings and different stages of the implementation. Technology provided by a technology provider should be useable and compatible in terms of system integration (Työppönen, 2018).

Design thinking has typically had a role in intervention development, while implementation of a solution into practice has not necessarily been addressed (Altman et al., 2018). While in the present study it is not a question of implementation in the sense of implementing a solution developed by the researchers, lessons learned from previous design thinking studies are to be noted. Altman et al. (2018), reviewing previous studies, emphasized that despite co-creation and field tests of a solution, there may be a lot of criticism and scepticism outside of actual pilot units. It is therefore essential to truly involve stakeholders to develop a user-centred process also for the implementation of a design thinking innovation. The context of the setting and users must be understood when both developing and implementing an intervention using a design thinking approach, and significant time and energy from stakeholders are needed.

Johansson-Pajala et al. (2019) examined appropriate circumstances for implementation of welfare technology, especially robotics in elderly care environments. Central issues are, according to them, convincing professional caregivers and multi-faceted collaboration. Caregivers need to be convinced about the usefulness and functionality of welfare technology and have a positive attitude towards technology use. They should see welfare technology use as part of care work. Some caregivers could be trained specifically for welfare technology use and then provide support to colleagues and older users. Collaboration should include all parties in the decision-making chain, but also older people themselves.

Cresswell et al. (2013) presented ten points to be taken into account when implementing technology in care services. Their model departs from clarifying what problem or problems are to be solved with the help of technology. After that, mutual understanding needs to be constructed and different options considered. These are included as basic elements in design thinking approaches (Altman et al., 2018; Ahookire et al., 2020, see also Figure 1). Good planning that does not overlook the infrastructure is important. Proper training for personnel should be emphasized in the planning. Progress needs to be monitored. Cresswell et al. (2013) list four stages in the lifespan of technology: need for change, selection of system, planning and maintenance and assessment.

Implementation requires work after the initial installation phase. Using a new service or device and giving feedback on it must be learned. Adoption and development of a new mode of operation belong in this phase. If the users are not convinced about the usefulness of a new service or product, they easily return to the old way of doing things. If the organization does not support the change, it is even easier to end up doing this (Hyppönen and Valkeakari, 2009). All these issues are points for both design thinking in general and, notably, information design. Organizational change may also cause resistance to change. The implementation may be deliberately hampered. Technology users assess personally if the change is negative or positive from their point of view. Resistance may manifest itself in many ways. Use of the device or service may be avoided; errors may be caused, and as time goes by, the object of resistance can change from the device to its meanings and become personalized in the supervisor. Resistance can be mitigated by providing sufficient support to both supervisors and users, committing all to the process and ensuring high-quality planning (Holm, 2013). Information design again plays a key role. As noted by Passini (1999), developers need to know how to solve problems so that the information can be brought forth in an appropriate way. Horn's (1999) definition of information design is apposite: "information design is defined as the art and science of preparing information so that it can be used by human beings with efficiency and effectiveness". Convincing users about usefulness, supporting change and mitigating resistance all require the use of information. Horn (1999) further noted that the values distinguishing information design from other kinds of design are efficiency and effectiveness at accomplishing the communicative purpose.

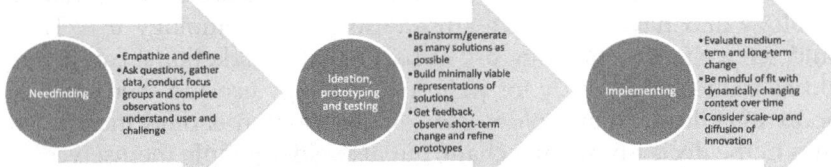

Figure 1. Design thinking process in healthcare (adapted from Altman et al., 2018).

Figure 2. Zora robot (Photograph by Satu Pekkarinen).

The purpose may be related to actual developing of comprehensible documents, designing easy human-computer interfaces, enabling people to find their way in virtual spaces or implementing and using welfare technology, such as in this study.

Challenges of implementation were also unveiled in a study on the implementation phase of a Zora robot (see Figure 2) in Finnish elderly care (Melkas et al., 2020b). Zora is a small humanoid robot designed for rehabilitation and recreation in social and healthcare contexts. Various challenges were brought up concerning the personnel, work plans and work atmosphere. There was too little time to learn to use the robot. Tensions arose between robot users and non-users, which led to changes in the work atmosphere. The use of the robot was felt to take too much time from hasty care work. Melkas et al. (2020b) concluded that orientation to technology use should contain issues related to time management and division of labour, in addition to technical details. All this necessitates good information design.

Orientation to welfare technology use

A vital part of the implementation process is orientation to the use of a device or service. As very practical activities and processes, orientation has often been overlooked in research. In the research project ORIENT (www.robotorientation.eu), which focussed on care robot orientation, the concept was defined as follows: "We perceive care robot orientation as the continuous co-creative process of introduction to technology use and its familiarisation, involving the learning of multi-faceted knowledge and skills for effective use" (Melkas et al., 2020a). This definition can also be used as a broader definition of orientation.

Orientation is highly affected by the type of device in question. Therefore, one clear model for orientation to welfare technology use is very difficult to create. However, orientation to technology tends to have certain similar features. Good orientation can contribute to overcoming or avoiding some of the implementation challenges discussed in the previous subsection. In addition, timely and adequate orientation has been found to be the most important means of reducing fears and negative impacts (Raappana and Melkas, 2009). This subsection briefly summarizes various issues to consider regarding orientation on the basis of earlier literature.

When planning an orientation process, it is good to consider the following questions: who, what, when, how and to whom (Eklund, 2018). Johansson-Pajala et al. brought up the same but placed 'when' under 'how' (Johansson-Pajala et al., 2019). Melkas et al. (2020a) included the 'why' question. 'Who' refers to the instructor, the person responsible for giving the orientation. If there are several instructors, the division of responsibilities must be clear. In addition, when appointing an instructor, it is important to ensure that resources are sufficient, both in terms of time and taking care of other tasks. 'What' refers to the topic(s) to be introduced. Different checklists may be handy because they increase the homogeneity of the orientation content between sessions. 'When' should be used to reflect on what should be said at which stage. Here, it is wise to evaluate which order and duration would best support the learning of the receiver. 'How' refers to the form of orientation that will be provided. This is an important part of ensuring effective and meaningful learning. 'To whom', on the other hand, refers to the receiver. How can individual characteristics be taken into account in the orientation? What are the receiver's wishes regarding the orientation? The essential questions are shown in Figure 3 (see also Eklund, 2018; Johansson-Pajala et al., 2019; Melkas et al., 2020a).

There are several forms of orientation: various training sessions, learning by doing, discussions with the instructor (giver of orientation), group work, independent study, online courses, webinars and videos and games (Eklund, 2018). The use of the device can be introduced, for example, by arranging a training session for the user or users by the device or system manufacturer. This training can be arranged either so that the trainer is in the same space as the trainee or by means of remote connections. One option for distance learning is to provide orientation

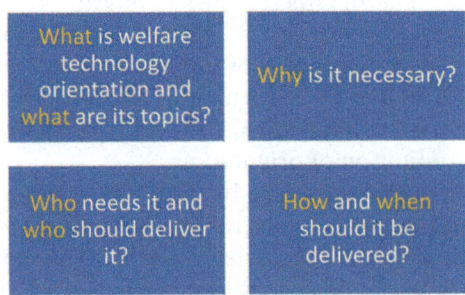

Figure 3. Planning an orientation process: The essential questions.

via a video call. The duration and extent of such orientation needs to be determined by the requirements of the device or system. One orientation option is peer training provided by another user. In this case, the instructor has received user training from either the device or system manufacturer or another user or has studied with the help of the operating instructions. The third option is that the user does not receive personal instruction but learns how to use the device through the user manual, a video or an online course (Melkas et al., 2020a).

As with implementation, there are challenges in orientation. For example, the short training traditionally provided by a device or system manufacturer in the early stages of technology implementation has been found to be poorly responsive to the needs of those who are supposed to adopt the use of the device. Also, the importance of continuous orientation is often not recognized at managerial levels of organizations (Raappana and Melkas, 2009). Taking care of the following things can help to make orientation successful: the management must understand the importance of orientation and be committed to its development; orientation must be carefully planned and adequately resourced; the timing of the orientation plan must be realistic; orientation should have clear goals and expectations; the needs and wishes of individuals in carrying out orientation should be taken into account as far as possible (Eklund, 2018).

Methods

This study is concerned with information design from the perspectives of education and teaching and construction of production as framed by Pettersson (1998). Graphic design is mainly beyond the scope of this study (see the arrow in Figure 4 showing the study focus).

Following Pettersson's categorizations, this study concentrates (1) on the communication processes as such—encompassing senders, receivers (also in double roles) and representation (instead of one of these) and (2) on the social context—the entire communications situation, i.e., senders

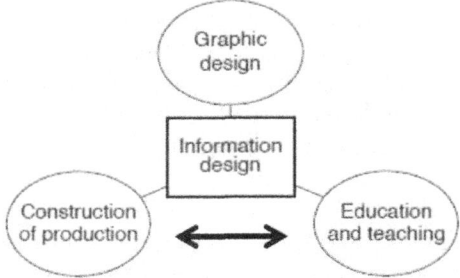

Figure 4. Perspectives of information design addressed (adapted from Pettersson, 1998, p. 28).

and their intentions for the verbo-visual message and receivers and their circumstances. The study mainly represents user research that forms part of the process according to Dyson's (2017) categorization of information design research methods.

This study seeks to distinguish and disaggregate needs for information design by utilizing Wang and Strong's comprehensive framework of information quality attributes and their definitions (Table 2) in the analysis of the empirical data. By doing this, this study contributes to understanding the role of information design in the design of services where welfare technology is to be used and proactively finding ways to enhance well-informed welfare technology use in society.

The study concentrates on the implementation processes of welfare technology in social and healthcare services in Finland, and the different micro-level users, that is, older end-users, professional caregivers and care managers. For the purposes of this study, they are categorized in groups, although that kind of approach has been rightly criticized by Eriksson (2016), as every individual has different needs.

End-users refer here to persons using devices at home. Implementation and orientation are examined via a number of welfare technologies. The whole of the large sphere of welfare technology cannot be examined within one study. This is a qualitative study, and the data collection was done with the help of written questionnaires and individual and group interviews. Data was also collected in workshops and via user pilots. The data collection methods of interviews, surveys, observation and literature studies are in line with methods for information design research listed by Dyson (2017) and Stahl-Timmins (2017). The research approach is inductive, creating broader meaning from individual observations. The study concludes with recommendations on what should be taken into account in information design when implementing welfare technology in the field of social and healthcare services from the perspectives of older people (clients), professional caregivers and care managers.

The implementation process of end-users (home use) was examined with the help of a robotic spoon pilot in which two people tested the product in their daily lives. The implementation process of professional caregivers was studied by interviewing caregivers in two work communities. All the perspectives were, in addition, approached in technology workshops that were organized with similar contents for all three groups—the end-users, professional caregivers and care managers. There are thus three datasets (in the order collected): one containing the interviews of the caregivers, another containing the data collected at the technology workshops, and a third containing the robotic spoon pilot data. There were altogether 23 participants, and five welfare technologies were addressed. The datasets and the data analysis are described in greater detail in Appendix 1.

Table 2. Information quality attributes and their definitions (source: Wang and Strong, 1996).

Information quality attribute	Definition
Believability	The extent to which information is accepted or regarded as true, real and credible.
Value added	The extent to which information is beneficial and provides advantages from its use.
Relevancy	The extent to which information is applicable and helpful for the task at hand.
Accuracy	The extent to which information is correct, reliable and certified free of error.
Interpretability	The extent to which information is in appropriate language and units and the information definitions are clear.
Ease of understanding	The extent to which information is clear, without ambiguity and easily comprehended.
Accessibility	The extent to which information is available or easily and quickly retrievable.
Objectivity	The extent to which information is unbiased (unprejudiced) and impartial.
Timeliness	The extent to which the age of the information is appropriate for the task at hand.
Completeness	The extent to which information is of sufficient breadth, depth and scope for the task at hand.
Traceability	The extent to which information is well documented, verifiable and easily attributed to a source.
Reputation	The extent to which information is trusted or highly regarded in terms of its source or content.
Consistent representation	The extent to which information is always presented in the same format and is compatible with previous information.
Cost-effectiveness	The extent to which the cost of collecting appropriate information is reasonable.
Ease of operation	The extent to which information is easily managed and manipulated (i.e., updated, moved, aggregated, reproduced, customized).
Variety of information and information sources	The extent to which information is available from several differing information sources.
Concise representation	The extent to which information is compactly represented without being overwhelming (i.e., brief in presentation, yet complete and to the point).
Access security	The extent to which access to information can be restricted and hence kept secure.
Appropriate amount of information	The extent to which the quantity or volume of available information is appropriate.
Flexibility	The extent to which information is expandable, adaptable and easily applied to other needs.

Welfare technology used in the study

In this study, welfare technology refers to various types of technology brought to social and healthcare services, such as therapy robots, exoskeletons and other assistive devices. In this study, different devices were selected that reflect the wide sphere of welfare technology. There were both devices for end-users and devices for caregivers. The devices are briefly introduced in Table 3.

Table 3. The devices used in the study.

Gyenno spoon is an assistive device that helps people with hand tremors and eating without stress. It has a 360-degree stabilization solution, offsetting 85% of unwanted tremors from the intended hand movement. It has an intelligent, high-speed servo control system that provides fast and accurate stabilization. It automatically distinguishes between intentional hand motion and unconscious tremors, only detecting unconscious tremors (www.gyenno.com). The spoon is shown in Photo 1. Experiences of use were studied in a thesis recently (Laine and Maunila, 2019). The results emphasized the need to test the spoon before acquiring it. Such testing would enable a determination of whether the spoon truly suits the user.	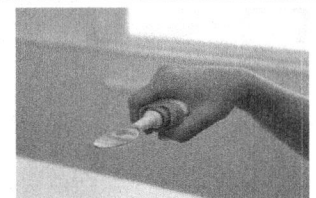 **Photo 1.** Gyenno spoon (photo: Päivi Tommola, 2019).
Joy for All cat (Photo 2) was designed to bring comfort, companionship and fun to older people. It resembles a real cat with a soft coat and feline movements and sounds (https://joyforall.com). The cat has sensors that recognize movement and touch, so it reacts by meowing and purring when touched or moved. The cat runs on batteries. In this study, the cat replaced JustoCat, which is professionally designed for therapeutic use, due to the practical arrangements of the study. In a previous study on JustoCat, it was found to have positive effects on the well-being and quality of life of memory-impaired persons due to its calming impact (Gustafsson et al., 2015).	 **Photo 2.** Joy for All companion cat (photo: Päivi Tommola, 2019).

Table 3 contd. ...

...Table 3 contd.

Laevo exoskeleton is a passive hip exoskeleton. It is designed to reduce the load on the lower back in bending situations. It supports the lower back during bending forward, while bending, when returning to an upright position and during lifting. The device operates on gas spring technology. Photo 3 shows the Laevo vest properly worn by a person. It is designed for indoor use only. It must not be worn by a person with a pacemaker or breast implants or whose axillary lymph node has been removed. (https://exoskeletonreport.com/product/laevo/).	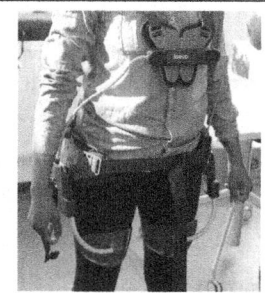 **Photo 3.** Laevo exoskeleton (photo: Päivi Tommola, 2019).
Oculus VR glasses Oculus Go is a portable and independent virtual glass (VR) system (Photo 4). VR glasses make it possible to watch virtual reality content, i.e., the users feel they are in 3D reality. In addition, the glasses have built-in surround controls, so no separate headphones are needed. This device is battery operated (Facebook Technologies, 2021). There are various free and paid software programs for VR glasses. In this study, VR glasses were used to display a waterfall landscape for relaxation.	 **Photo 4.** Oculus VR glasses (photo: Päivi Tommola, 2019).
Yeti tablet is a giant, multi-purpose tablet. The Yeti tablet's large screen makes it easier for the visually impaired and people with reduced mobility to use it. With the Yeti tablet, a person can, among other things, play games, watch programs from streaming services and read the news. It can be used for activities alone or in a group. In Photo 5, the Yeti tablet is in use, and its size range can be seen from the image (Kuori Oy, 2021). In this study, the use of the Yeti tablet was related to engagement and rehabilitation of clients in company-operated care facilities, for example, through various games.	 **Photo 5.** Yeti tablet (Kuori Oy, 2021).

Needs for implementation and orientation

The results concerning needs for implementation and orientation related to welfare technology are presented in this section, starting from the older *end-users* and moving on to the *caregivers* and *care managers*.

End-users at home

The end-users wished that there would be mutual understanding in orientation, and they wanted to be sure of having an experience of managing the device after the orientation. The participants in the older people's workshop noted, for example,

> So they say, ... if you go and ask, they just say that 'this is easy to use ... like this and this'. So you should bring understanding to older people.
>
> They say very quickly that 'yes, you can, you just do like this-and-this'.

The importance of testing and using the device oneself during the orientation was expressed by one end-user as follows:

> You have to do it so that we understand. So that we also get to do it; that they [givers of orientation] stay by us and oversee.

The support person after the orientation was felt to be important because then it would be possible to ask for help when having problems. The end-user group also hoped that technical language would be avoided or clearly 'translated into Finnish' during the orientation. One end-user commented,

> ... if there is some technical vocabulary that this person uses who brings the device ... that we talk with such words that we understand them, the practical words, what they mean, so that we really get it.

Online training was not preferred by this group. At least the beginning of the orientation should be hands-on training:

> It took a long time for me to learn to use this tablet because I had to do it all by myself.

Older people as a user group have been left in a weaker position in the design of products (or services). There is product and service design that takes them into consideration, but on the basis of this study, that kind of design should clearly be increased. Another development idea might be older people's own service channel that would provide support for taking the most usual devices, such as tablets, into use. Non-governmental organizations and municipalities may offer various types of courses for using such devices, but the threshold to participate may be high if the person already has many poor deployments behind them.

The main point in the implementation process is that the end-users can test and use the device themselves. The end-users must be listened to and their perspectives taken into account. The orientation should be based on mutual understanding between the giver and receiver of orientation, and the end-users should experience being able to manage the use of the device. If the use of technical terms cannot be avoided altogether, it needs to be ensured in the orientation that the user understands the terms. The participants of the technology workshop also felt that it is important to have a support person after the orientation. Online training was not considered appropriate; at least the first orientation should be conducted 'holding hands'.

As to the attributes of high-quality information, seven attributes were identified as particularly necessary for the end-users, to be taken into account in information design: believability, relevancy, interpretability, ease of understanding, accessibility, objectivity and reputation.

Caregivers

Professional caregivers' views and needs for implementation and orientation were collected at a technology workshop and in interviews (see Appendix 1). In the group of professional caregivers, personal guidance for use was desired, not just the reading of a manual. However, this was perceived to be device-dependent, i.e., for easy-to-use devices, a simple introduction to the manual may suffice. It was generally felt to be vital that orientation be available:

> *Just so that someone with practical knowledge would come along, that not just papers are brought ... 'read these, and then learn to use'. The kind of personal guidance you probably need, during the first times, for the person who will use the device with clients. Everything except the cat requires this [reference to the Joy for All companion cat]. With the cat, it's enough to have the instructions on paper, 'switch on here and take it in your arms'.*

In addition, it was hoped that during the guidance for the use of the device, it would be explained why the device is being put into use, i.e., the benefits of using it would be justified. The benefits included, for example, the perspective of well-being at work and better customer satisfaction:

> *In my opinion, all kinds of well-being at work, be it a supervisor or a resident or a caregiver, so yes, all means of maintaining well-being at work must be brought up. Of course, like asking about your holiday wishes, just as much, wishes about technology should go hand in hand, if it [technology] just makes work easier or the life of a resident better somehow. Why not?*

> *If you're bringing something new, hardware or something like that, you really need to get that orientation, and some reasons you are given for why something is brought to the workplace. Not so that something is just brought in, and it is said that 'take this into use with clients'. It will definitely not be used. You also need to then... maybe tell just those benefits.*

Consideration of one's own attitude emerged as an important aspect in this group. It was felt that a bad attitude does not get clients excited, but enthusiasm inspires clients:

> *Always bringing in something new, unless you're very enthusiastic... you can't get anything through. As an instructor, you really need to stand behind that. That's when you get them [clients] to participate, I've really noticed. If I think it is great, well, they will immediately come along and think it is great. That is how we start to bring those new things inside.*

It was also considered essential that the use of the device be a matter for the entire staff and not remain the responsibility of one person. Managers' support was felt to be needed because procurement is the responsibility of management. In addition, it was hoped that procurement and experimentation would be caregiver-driven, too. Thus, participation in the use of technologies was thought to be more secure when user experiments or purchases are made at the request of the caregivers:

> *[It is important] that the experiments also depart from the caregiver level. Then it [the change] goes through.*

Distance training was also seen as a possible option:

> *For example, that Laevo could go with Skype quite well. You get the device, and then there is someone else telling you, if you have problems with some adjustment, how you tighten it. ... It is always nice to see people face-to-face, but yes, that would certainly succeed.*

Group interview with caregivers in a private care facility (Unit 1)
The interview with Unit 1 (see Appendix 1 for details) highlighted good experiences with technology use. The unit's personnel were interested in new technology. They felt that only part of the opportunities are utilized now. In their work, they found that the use of technologies clearly brings added value to clients. They also felt that they had gained added value to their work by using different technologies. Technologies guided the order in which their tasks were performed.

When asked about their needs and dreams of services utilizing technology, it became clear that caregivers experience a lack of information about the opportunities available. Among other things, technology is expected to help create a sense of safety and companionship for the clients. In addition, a voice or video connection to the clients' rooms, connected to the call system, was hoped for. This was intended as a solution for reassuring the clients in situations in which they cannot be visited immediately. The personnel thought this would increase the clients' sense of safety and security. In addition, it would also increase service quality as it would allow time to react more quickly to the clients' concerns. The feeling of insufficiency of caregivers in call situations could also be reduced.

In the implementation of new technology, the general practice in the unit is that the introduction to the device is initially done by the party selling it. After that, the caregivers train each other, i.e., the person who has already received the orientation can give orientation to another person. Manuals are read as needed. The personnel find this a good practice. The interviewees emphasized their own responsibility to be informed and receive orientation and keep up with the changes:

> *And you have to take responsibility for that, too, yourself—you can't leave it so that you expect someone to tell you, but you also have to take responsibility for finding out if something new has come. It is a reciprocal responsibility, that is how it works.*

The interviewees knew that there is a wealth of information available. However, they found it challenging to find a suitable time for information retrieval. All working time was reported to be spent on care work in the field:

> *Information is available today. But it is perhaps more an issue of where the time can be found to do so... to do that information search. After all, ...all of our employees are pretty much there, right in the field, doing care.*

The personnel feel that the clients 'absorb' orientation to the use of devices well. After just a few sessions of orientation, the clients used the devices independently. It was hoped that, in the future, the clients would have better technological knowledge and a wider opportunity to participate in device purchases. Currently, procurement in the unit is decided by the senior management. The personnel want to create the best possible experiences for the clients, so being involved in the procurement phase would be desirable from that point of view as well:

> *...there should first be enough information, and understanding of different opportunities available, and then the residents should be involved in choosing. Because it is them you should think about.*

Group interview with caregivers in a private care facility (Unit 2)

In Unit 2 (see Appendix 1 for details), the new technology was perceived as a very useful aid at work. On a scale of 1–5 (1 = I don't feel like I'm getting help from technology for my job, 5 = I feel like I'm getting a lot of help from new technology for my job), the average was 4.7. Time use, enthusiasm and motivation were seen as challenges and pitfalls for orientation. Sufficient time is needed for familiarization per device and service so that people can get the most out of the technology. Utilizing technology can even provide more time for care work. One interviewee commented on the orientation challenges:

> *It takes time. It really takes a lot of time ... we had the good fortune that we could all be here at the same time, we went through and got acquainted with all those devices at the same time.*

The importance of orientation as a time saver for the future was also brought up. Without adequate orientation, important functions of the device or service may go unnoticed. With good familiarity, the functions of the device or service are clear and easy to use. The interviewees gave an example about using the oven, how they had found a timing function in the oven that made it easier to make morning porridge, among other things. Thus, with proper use, time can be saved that can be redirected to care work.

The motivation of the caregivers was also felt to be essential. One interviewee commented,

And this one needs that enthusiasm and motivation. If someone thinks that well, those others use it, I don't need to be able, I don't need to know. But it's good for everyone to be interested and want to know and be part of it.

If an employee is not enthusiastic about using the technology, it is difficult to get the client inspired. In general, orientation to technology (given in previous workplaces) was not felt to be very good.

Unit 2 has a Yeti tablet. The personnel had been instructed in the use of the device through Skype training. The instructor provided training over a video call, and the caregivers learned by doing it themselves. The orientation could also have been carried out in such a way that the instructor would have come on site to guide use. It appears from the interview that some of the caregivers would have needed an on-site orientation, while some were satisfied with distance learning. One interviewee noted,

We had to do it ourselves while he instructed on the phone. It was all right, but if he had been here, he might have shown himself. Had it been face-to-face, would he have shown more of the use...? But you learn better by doing than by watching.

When giving orientation to clients, it was seen as a pitfall that the wrong type of devices or services are chosen for them due to lack of information. In addition, a situation in which a client is left to use a device or service alone too early is considered a pitfall:

After all, there's a lot [applications on the Yeti tablet] that, yes, it certainly requires each of the residents in the beginning to be guided 'hand in hand', with one of us present and doing things together—until the residents can use it by themselves.

Lessons learned from caregivers' needs

The caregivers raised several important issues related to implementing and becoming oriented to welfare technology. Attitude was mentioned in both the interviews and the workshop. When caregivers are enthusiastic, they inspire clients to try to use the devices. It was also seen as important that procurement and implementation have managers' support.

Managers define the resources according to which the implementation of and orientation to devices and services are carried out. It was also hoped that procurement and trial pilots could be more caregiver driven. The commitment of all personnel to the use of the devices was also felt to be important. In this wáy, pilots are not the sole responsibility of one person: the entire staff assumes responsibility.

The orientation was hoped to include personalized guidance adapted to the complexity of using the device, not just reading the instruction manual. The justification for the implementation of the device was also mentioned: it must be clear why the device is being put into use. Thus, for example, the benefits of using the device should be justified. In this context, well-being at work and its promotion were seen as an important benefit and an issue to be considered. Perhaps the most important aspect was that orientation is provided at all, i.e., that learning to use a device is not entirely the employee's own responsibility. Distance learning was perceived as one good form of orientation, depending on the device.

As to the attributes of high-quality information, seven attributes were identified as particularly necessary for the caregivers: value added, relevancy, ease of understanding, accessibility, completeness, concise representation and appropriate amount of information. These attributes to be taken into account in information design were partly similar to those identified from the end-users' dataset.

Care managers

Committing users to the use of new technology was perceived as challenging by care managers at their technology workshop (see Appendix 1 for details). It is considered pivotal to enable employees to find personal benefits to the new technology. The use of the device must be guided by a clear idea of why it is used. Every device should have a goal that would identify why it is being used and what benefits it will bring:

> *In my opinion, for this device [the Joy for All cat], and for so many other devices, the most important orientation would be the joint discussion about what we are aiming for with this device. That is, it is not so much about the instructions of using that device. Those are essential, too, because if they are not given, the device will not be used. However in addition, a joint discussion is needed about things like what this device is, why it is in use, what we are trying to reach with this device, with whom it works, and with whom it perhaps doesn't work.*

If the aim of using the device is to reassure the customer, this should be clearly stated during the orientation to the device. It is also important to go through together what kinds of clients can benefit from the device. The idea that the implementation and use of devices should be considered as part of the whole of care procedure also came up. In this way, the utilization of technology does not remain an unsteady part of care work.

Attention was also paid to the storage and maintenance of the device. Consistent instructions were felt to be necessary, so that every member of the work community knows how to act correctly with the situation. This was seen as part of enabling safe usage. Appropriate common use practices and storage rules need to be agreed on (instructions for handling, storing and transferring the device—for whom and when). These guidelines can save the owner from potential breakage or theft of expensive devices. Other concerns were also noted:

Of course the hygiene, too; it is, especially if a larger group uses a device, quite important for all users to know about ... if something needs to be taken into account, and what you can use, and how you can maintain the cleanliness and hygiene.

And if it breaks, or, goes out of order ... that is always a challenge; what to do now, where do you take it or who will repair it.... What's wrong with it... With so many devices you really have no idea, you don't know what to do in the first place, and where, who knows, and whom you are supposed to contact.

The choice of the right end-user was seen as important in terms of guidance. This issue was also perceived as a safety factor. Technology was felt in part to even impose limitations on memory-impaired users in particular. The participants thought that a limiting factor is, for example, that one does not remember which part of a device should be put in the washing machine and which part has to be wiped clean. That is, cleaning the device may limit use by memory-impaired users.

There were also a lot of opportunities and positive features found in the use of technology. Among other things, the use of technology was felt to increase the clients' right to self-determination and independence, as well as to promote functional abilities. With technology, caring for mental well-being was seen as one possibility. For example, refreshing activities could be practiced even by bedridden patients:

That is definitely a type of technology that brings hope [Gyenno smart spoon], and if it works well ... when used, it is really good. It is about increasing independence and self-determination, promoting one's own ability to act. Assuming it does work, it's good.

But I experienced it so, as discussed, that it [VR glasses] can make it possible to restore things that you may have to give up due to age or limitations. I am thinking, for example, of cultural issues, art exhibitions, music concerts. Of course, some may like to go there again and again ... to the roller coaster and experience the charm of speed. But it can also be used to seek out the relaxation on the beach or a musical experience, which may not be possible otherwise. Yes, I see enormous opportunities.

The participants wanted to pay attention to the technology adoption process, too. One participant commented,

Somehow attention should be paid to the process of technology adoption, not just orientation to use. However, it is a new way of working.

The adoption of technology was perceived as possible when the concrete benefits are widely highlighted. Both clients and caregivers can gain benefits. For example, highlighting aspects of well-being at work and occupational safety help visualize these concrete benefits:

Very often, technology is also thought of as just aimed at playing games or bringing joy, or something like this, but you should see it as broader. So that while it brings joy, it can also bring some tangible benefit. Or, in that exoskeleton, for instance, the benefit is in occupational safety and well-being at work, as well. In a way, you should see it [technology] as broadly as possible; not just that pleasure and enjoyment, and variety and stimulation, but also that benefit aspect, for everyone. And again, not just for the older people, but also for the employees.

Care managers saw distance learning as a good option depending on the device in question. They themselves had not received training in the use of devices through distance learning. They had, however, been introduced to the use of computer programs as distance learning.

As to the attributes of high-quality information, eight attributes were identified as particularly necessary for the care managers: value added, relevancy, interpretability, ease of understanding, accessibility, completeness, concise representation and appropriate amount of information. These attributes to be taken into account in information design were partly similar to those identified from the end-users' dataset and the caregivers' dataset.

Information-design-related lessons learned

This section summarizes essential issues related to implementation and orientation of welfare technology, discusses central information quality attributes and shows what was done to bring the lessons learned to better use in practice.

Implementation and orientation

A number of essential issues to be taken into account when implementing welfare technology in the field of social and healthcare services were identified in this study. Figure 5 provides a summary.

188 *Different Perspectives in Design Thinking*

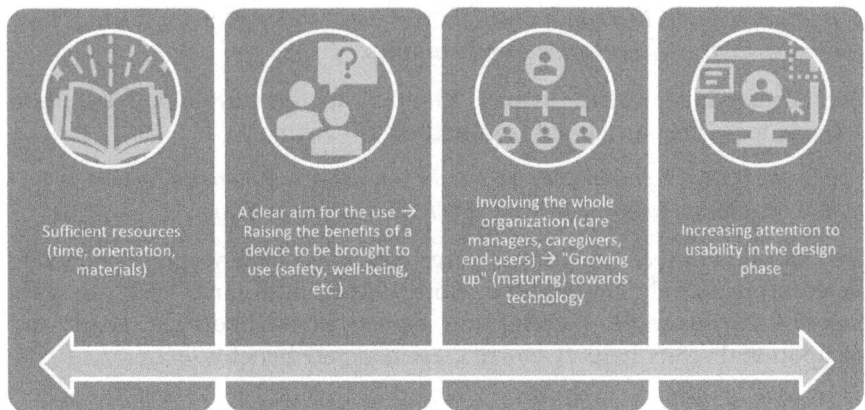

Figure 5. Issues to consider in the implementation of welfare technology, to be addressed via information design.

All of these issues should be addressed via information design; this concerns different methods related to form, function, management actions and engagement. In practice, they range from graphic design of materials to refining the contents, engaging senders and receivers of information—as well as those in double roles—and elicitation and provision of information to those who design devices (and services more generally than within one unit, for example). The above-mentioned issues also emphasize the importance of being holistically interested in the communication processes and the social context and examining them as dynamic parts of the whole. Many of them are related to Horn's (1999) guidelines for information design: (1) To develop documents that are comprehensible, rapidly and accurately retrievable and easy to translate into effective action; (2) To design interactions with equipment that is easy, natural and as pleasant as possible. This involves solving many problems in the design of the human-computer interface; (3) To enable people to find their way in three-dimensional space with comfort and ease—especially urban space, but also, given recent developments, virtual space. In this case, it is not a question of an urban space or a virtual space, but rather a physical space enriched with technologies in the context in question—either a home or a workplace.

When planning implementation, sufficient resources must be set aside to enable a smooth start. These resources refer to time, orientation and materials. Users must be guaranteed high-quality and sufficiently long-term training. Overall, sufficient time must be taken for the implementation process. It should cover communication, orientation and, most importantly, allow the fact that, initially, using the new device will take more time. Whenever possible, this should be taken into account in human resources so that each user has time to use the device in their work. This would also prevent the device from being left unused due to incompetence. Time should also be set aside for making good orientation materials. From a financial point of view, it should be taken into account that when purchasing a device, the price of the device is not the only cost: money must also be budgeted for the implementation process.

The results confirmed that there must be a clear goal for the use of the device. Why is it put to use? What is hoped to be achieved with its use? Goals can be related to a single thing or bring together many things. For example, with the help of a therapeutic cat intended to reassure the client, a feeling of calmness can be achieved (Gustafsson et al., 2015, on JustoCat). Calming the client also guarantees a safer working environment for the caregiver. An aggressive client can possibly be reassured by drawing her attention to the device. The aim of use can therefore be both to calm the clients and to increase the occupational safety of the personnel.

Involving the entire organization in the implementation process was emphasized. When all employees throughout the organization are involved in the process, there is an increase in 'growing up' towards technology at the same time. Growing up towards technology means getting used to implementation processes and the use of different technologies. Every employee in the organization has his own role in enabling a smooth implementation process. Managers enable procurement and resources from the financial side and, through their management, guide the personnel towards utilization of technology. A well-designed implementation process, for which the goal of technology use is clearly defined, plays an important role (see also the other points in Figure 4 as well as Table 4). If managers do not truly support the implementation process, it will be more difficult to carry out. For caregivers and users, considering their own attitudes is essential. Any previous bad experiences should be forgotten, and people should feel free to try using new devices.

The more user views are taken into account in the design phase of the device, the easier it becomes to use the finalized device. Consideration of usability in the design phase is an important part of the whole process. It is also necessary to consider different user groups in the design. When designing devices for the social and healthcare fields, there are often at least two user groups, caregivers and end-users (sometimes also their relatives). Consideration of usability can also facilitate the implementation process and the growing up towards technology. When the users of the device feel that they have been taken into account and listened to in the design of the device, the final taking into use is smoother to implement. Table 4 further summarizes the different groups' needs for implementation and orientation.

Accounting for the needs of different groups can be made possible by good planning of the implementation process as part of service design. This notion was also confirmed in the study of Johansson-Pajala et al.

Table 4. Different groups' needs for implementation and orientation.

Care managers	Caregivers	End-users
Challenging to commit users to using new technology → Finding one's personal benefit needs to be enabled for caregivers	Personal orientation to the user, not just reading of a manual (depending on the device)	End-users must be listened to and their views taken into account
A clear goal for why technology is used	Orientation emphasizing why a device is brought to use and its aim, e.g., well-being at work	→ Mutual understanding in orientation; users' experience that they can manage use
→ Aim, why is it used, what benefits does it bring	→ Communicating benefits Paying attention to one's own attitude (caregivers' enthusiasm makes clients enthusiastic)	End-users' opportunity to try using the device themselves during orientation
Storage and maintenance → Consistent practices → Safe use	Not just one person in charge of use but the whole staff	Avoiding technical vocabulary or providing its clear interpretation during orientation → No unclear words/ concepts
The right technologies directed at the right people → Safe use	Managers' support (funding, caregiver-driven pilots) Distance training possible	A support person to give guidance after orientation
Attention to adoption process of technology → Raising concrete benefits widely (occupational safety, well-being at work)		Distance training not an option → First orientation, at least, to be conducted together, 'holding hands'
Distance training possible, depending on the device		

(2019). As shown in Table 4, the purpose of using the devices comes forth in the needs of both managers and caregivers. Cresswell et al. (2013) also emphasized this in their study. Jauhiainen and Sihvo (2015) highlighted both the description of the need and the evaluation of effectiveness of technology. In addition, Table 4 shows the need expressed by caregivers to ensure that technology use is the responsibility of all personnel and that there is management support behind it. This importance of the organization is also emphasized by Työppönen (2018).

The needs for orientation are most pronounced in the group of end-users, but the caregivers were also hoping for it. It was felt to be important that the user has a feeling that she is in control of the device during the orientation. Orientation must have a good plan behind it so that different goals and needs are taken into account and met. This was also highlighted by Eklund (2018). Distance learning was felt to be prospective, depending on the device; older end-users did not see it as an option, however.

Information quality attributes

On the basis of the analysis presented in this chapter, high-quality information was essential for welfare technology implementation. All the attributes of information quality were not central, however. Some of the attributes identified as particularly central per group were the same for all the groups, whereas there were also a few different emphases. Appendix 2 gives details on the results. Many of the lacking attributes were due to the context of the study; traceability and access security are attributes associated with different types of operations and information use.

Relevancy, ease of understanding and accessibility were pivotal for all three groups. On the other hand, believability, objectivity and reputation-related issues were emphasized by older people, not the other groups. The differences are natural due to the participants being either in work roles or roles as individuals commenting on their daily lives. In this study, the participants were representatives of their own groups and the grass-roots service level, but the more multi-sectoral and multi-professional networks we talk about—they are also relevant for welfare technology services and the social and healthcare sector—the more likely is the need for ease of understanding and interpretability.

By taking information quality attributes into close consideration in information design for implementation and orientation processes, meaningful and well-informed welfare technology use can be enhanced. In particular, the attributes marked as central in Appendix 2 should be paid attention to in such information and service design, and various plans could be tested against the attributes and their definitions.

In general, all actors in roles in which they can influence planning should pay close attention to ease of understanding, relevancy and accessibility of the information that they work with by, for example, using appropriate concepts and less complicated language than they might use with their own colleagues at their workplace, as well as providing an appropriate amount of information that is represented concisely (Jacobson, 1999). This kind of thinking is simple yet demanding and powerful. Implementation and orientation processes are—or should be—empowerment processes, and, as noted by Simlinger (2014) in the context of information design, high-quality information empowers people to attain goals. In the empowerment process, taking into account quality attributes may help in creating a joint understanding and commitment to collaborative action.

During the empowerment process, in group situations, such as the workshops in this study, sharing and co-creation of information becomes possible (see also Altman et al., 2018; Ahookire et al., 2020, on design thinking). Those who have negative experiences or attitudes towards welfare technology also get to hear other views and arguments to support their thinking and reflection. In group situations, 'warm-up' is needed in this topic as there is uncertainty around technology issues, and people may be inclined to think that only technology experts can have their say. It can indeed be challenging to engage non-interested people in co-creation of user information. Again, this is an information design issue that may have to do with, for example, how individual people's experiences of using technology in different contexts are brought up—with the help of design thinking approaches—and presented to make them seem believable and accessible. As noted by Passini (1999), respecting the receiver's information-processing characteristics is fundamental to transferring information. The ways people read and understand messages vary with the task and the individual; they are not the same when confronted with a set of instructions at work or when struggling to make sense of complex issues. Understanding the ways in which people process different kinds of information provides cues to the most suitable forms for presenting that information (Passini, 1999).

There are links to wider communication processes: how does the media or how do patient organizations perhaps engage in dissemination of welfare-technology-related information? How can advertising of individual products be avoided? How can different kinds of short- and long-term impacts be assessed and visualized? How is it possible to move from mere grains of information towards sufficient information reserves? Peer support and information from peers typically interest people in this case. There may be importers' or producers' experiences and posts on social media, but knowing how accurate and relevant that kind of information is for oneself, with one's own characteristics and needs, is difficult. Again, these issues highlight the need for good information design.

Visualization and use of the lessons learned

To bring the results of this analysis to better use in practice, discussion cards and an animation were produced for use in social and healthcare workplaces and in the education of future care professionals (see also Pettersson, 1998). These means were selected in line with Passini's (1999) view of information design, meaning communication by words, pictures, charts, graphs, maps, pictograms and cartoons, whether by conventional or electronic means. The animation (see an example of its contents in Figure 6) is in Finnish, but Swedish and English subtitles are available. The discussion cards are in Finnish (see Figure 7 for examples). The cards support people in pondering the use of welfare technology, what should be and can be done to make welfare technology serve us as meaningfully

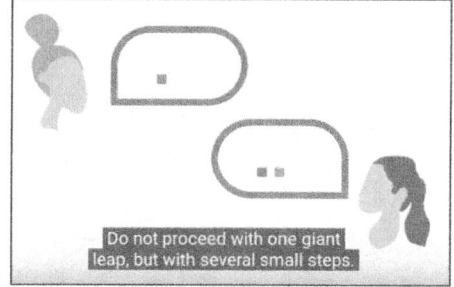

Figure 6. Examples of the animation (source: Melkas et al., 2020c).

as possible, now and in the future. The cards can be used when considering one's own and joint choices in the workplace or other contexts in which the technology is used. There are 15 cards for use together or individually to suit to the situation or context. The contents are related to the acquisition of welfare technology, orientation to it, its integration to daily activities, clients' services and well-being, assessment and the role of technology in developing services. They give ideas and guide thinking and discussion.

This study as well as the authors' previous studies have underlined that approaching and taking welfare technology into use resembles a problem-solving activity. A technological device is at least supposed to solve a problem in an individual's life or at work—for example, a problem related to safety and security, autonomy, loneliness, declining cognitive abilities, haste at work, ergonomic challenges or the like. On the other hand, selecting an appropriate device, learning how to use it, integrating it into daily life or work and getting used to it—these are also problem-solving activities. As noted by Passini (1999), "whenever people set goals and use information to attain those goals in novel conditions, they engage in a mental activity that can be conceptualized as problem solving. Providing information for problem solving is one of the major tasks of information design. Understanding how people solve problems provides designers with the criteria needed to determine what information is required and where or when it has to be accessible".

Various variables can affect information processing. Information overload may reduce the processing of information. There may be

Figure 7. Examples of the discussion cards (source: Hennala et al., 2020).

cultural, social and age differences that make responses to information more complex (Passini, 1999)—this is often the case with welfare technology used by older people. The discussion cards and the animation are mainly targeted to professional caregivers, present or future, with the aim of making the messages less complex and emphasizing information-processing differences between receivers (see also Raskin, 1999) and contextual variables. It is thus hoped that these human-centred and user-centred outputs would benefit the welfare technology users in question. Ahookire et al. (2020) emphasized that the design thinking process aims to promote action over individual thinking, and such action can include discussion and questioning, so they may also support general design thinking.

Conclusion

In this chapter, we have discussed research that provides information on the implementation and orientation of welfare technology in the social and healthcare services from three different perspectives. This information, including the three groups' perspectives, provides food for thought about welfare-technology-related design thinking, notably information design. Attributes for high-quality information needed by the three groups were identified—being partly different and similar. The results of the research can be utilized when planning implementation and orientation processes related to welfare technology's use in care services. The use of the results—with or without tangible outcomes such as discussion cards and animations—could ensure that the needs of end-users, caregivers and care managers are taken into account better in the future. The research approach appears usable also for other contexts and services. Future research could focus on studying larger groups of people, other devices and other countries. International joint research on this topic could contribute to creating international guidelines for information design for welfare technology use.

Acknowledgements

This research was supported by the Strategic Research Council (SRC) established within the Academy of Finland (project name: Robots and the Future of Welfare Services, ROSE, decision numbers 292980 and 314180). Support from the European Regional Development Fund and the Regional Council of Päijät-Häme is also gratefully acknowledged (project name: HyTeLab - Päijät-Häme's well-being technology innovation, test and development environment). The workshops were organized in collaboration with the ORIENT project (JPI MYBL; national funder: Academy of Finland, decision number 318837).

References

Ahookire, S., Plover, C., Frasso, R. and Ku, B. 2020. Health design thinking: An innovative approach in public health to defining problems and finding solutions. Front. Public Health, 8(459). https://doi.org/10.3389/fpubh.2020.00459.

Altman, M., Huang, T.T.K. and Breland, J.Y. 2018. Design Thinking in Health Care. Preventing Chronic Disease 15 (180128). http://dx.doi.org/10.5888/pcd15.180128externaliconhttp://dx.doi.org/10.5888/pcd15.180128external icon.

Black, A., Luna, P., Lund, O. and Walker, S. 2017. Information Design: Research and Practice. Taylor & Francis.

Bouma, H. 1998. Gerontechnology: Emerging technologies and their impact on aging in society. pp. 93–104. *In*: Graafmans, J., Taipale, V. and Charness, N. [eds.]. Gerontechnology: A Sustainable Investment in the Future. IOS, Amsterdam.
Chasanidou, D., Gasparini, A.A. and Lee, E. 2015. Design thinking methods and tools for innovation. pp. 12–23. *In*: Marcus, A. [ed.]. Design, User Experience, and Usability: Design Discourse. Lecture Notes in Computer Science, 9186. Springer. https://doi.org/10.1007/978-3-319-20886-2_2.
Cresswell, K., Bates, D. and Sheikh, A. 2013. Ten key considerations for the successful implementation and adoption of large-scale health information technology. Journal of the American Medical Informatics Association 20. https://doi.org/10.1136/amiajnl-2013-001684.
Dyson, M.C. 2017. Information design research methods. pp. 435–450. *In*: Black, A., Luna, P., Lund, O. and Walker, S. [eds.]. Information Design: Research and Practice. Taylor & Francis.
Easterby, R. and Zwaga, H. [eds.]. 1984. Information Design. The Design and Evaluation of Signs and Printed Material. John Wiley and Sons Ltd.
Eklund, A. 2018. Tervetuloa meille! Uuden työntekijän perehdytys [Welcome to Us! Orientation for a New Employee; in Finnish]. Grano Oy, Helsinki, Finland.
Eriksson, Y. 2016. Technologically mature, but with limited capabilities. pp. 3–12. *In*: Zhou, J. and Salvendy, G. [eds.]. Human Aspects of IT for the Aged Population. Design for Aging: Second International Conference, ITAP 2016, Toronto, ON, Canada, July 17–22, 2016, Proceedings, Part I. Lecture Notes in Computer Science Information Systems and Applications. Springer.
Facebook Technologies. 2021. Oculus Go. Features. https://www.oculus.com/go/features/.
Gammon, D., Strand, M. and Eng, L.S. 2014. Service users' perspectives in the design of an online tool for assisted self-help in mental health: A case study of implications. International Journal of Mental Health Systems, 8(1): 2.
Goodhue, D.L. 1995. Understanding user evaluations of information systems. Management Science, 41(12): 1827–1844.
Gustafsson, C., Svanberg, C. and Müllersdorf, M. 2015. Using a robotic cat in dementia care: A pilot study. Journal of Gerontological Nursing, 41. doi: 10.1186/s12913-016-1913-5.
Hennala, L., Pekkarinen, S., Melkas, H., Saurio, R., Rinkinen, S. et al. 2020. Hyvinvointiteknologian hyödyntäminen: keskustelukortit [Utilization of welfare technology: Discussion cards; in Finnish]. LUT Scientific and Expertise Publications: Oppimateriaalit - Lecture Notes 18. LUT University. Lahti, Finland. http://urn.fi/URN:ISBN:978-952-335-55-2.
Holm, T. 2013. Tietojärjestelmän käyttöönotto ja sen hyväksymiseen vaikuttavat tekijät [Implementation of an Information System and Factors Affecting its Acceptance; in Finnish]. University of Jyväskylä, Jyväskylä, Finland.
Horn, R.E. 1999. Information design: Emergence of a new profession. pp. 15–16. *In*: Jacobson, R. [eds.]. Information Design. The MIT Press. Cambridge, Massachusetts.
Hyppönen, H. and Valkeakari, S. 2019. Muutosvalmennus sähköisten palveluiden käyttöönoton tukena: Case Oulu omahoito [Change Training to Support Implementation of Electronic Services: Case Oulu Self-Care; in Finnish]. Report 34/2009. https://www.julkari.fi/bitstream/handle/10024/80100/51d94952-7fbe-49ad-b19f-cc2bb33480cc.pdf?sequence=1 (accessed 20 July 2019).
Jacobson, R. 1999. Information Design. The MIT Press. Cambridge, Massachusetts.
Jauhiainen, A. and Sihvo, P. 2015. Asiakaslähtöisten sähköisten terveyspalvelujen käyttöönotto – malli käyttöönotolle ja vaikuttavuuden arvioinnille [Implementation of User-Centred Electronic Health Services—A Model for Implementation and Evaluation of Effectiveness; in Finnish]. Finnish Journal of eHealth and eWelfare 4.
Johansson-Pajala, R.-M., Thommes, K., Hoppe, J.A., Tuisku, O., Hennala, L. et al. 2019. Care robot orientation: What, who and how? Potential users' perceptions. International Journal of Social Robotics. https://doi.org/10.1007/s12369-020-00619-y.
Kuori Oy. 2021. Yetitablet for Healthcare. https://yetitablet.com/healthcare.

Laine, C. and Maunila, A.-M. 2019. Käyttäjäkokemuksia käden vapinaa stabiloivasta Gyennoälylusikasta [User Experiences of Hand Tremor Stabilizing Gyenno Smart Spoon; in Finnish]. http://urn.fi/URN:NBN:fi:amk-2019053113831 (accessed 26 February 2020).
Lee, Y.W., Strong, D.M., Kahn, B.K. and Wang, R.Y. 2002. AIMQ: A methodology for information quality assessment. Information & Management, 40(2): 133–146.
Linton, J.D. 2002. Implementation Research: State of the Art and Future Directions. Technovation 22.
Melkas, H., Uotila, T. and Kallio, A. 2010. Information quality and absorptive capacity in service and product innovation processes. Interdisciplinary Journal of Information, Knowledge and Management, 5: 357–374.
Melkas, H. 2013. Innovative assistive technology in Finnish public elderly-care services: A focus on productivity. Work: A Journal of Prevention, Assessment and Rehabilitation, 46(1): 77–91.
Melkas, H., Gustafsson, C., Hennala, L., Pekkarinen, S., Tuisku, O. et al. 2020a. Care Robotics: Orientation Pathways for Users and the Society. Lappeenranta-Lahti University of Technology LUT, LUT Scientific and Expertise Publications, Tutkimusraportit – Research Reports 106. Lahti, Finland. https://www.robotorientation.eu/wp-content/uploads/2020/03/Orient-Care-robotics.pdf (accessed 26 June 2020).
Melkas, H., Hennala, L., Pekkarinen, S. and Kyrki, V. 2020b. Impacts of robot implementation on care personnel and clients in elderly-care institutions. International Journal of Medical Informatics, 134. https://doi.org/10.1016/j.ijmedinf.2019.104041.
Melkas, H., Pekkarinen, S., Hennala, L., Saurio, R., Rinkinen, S. et al. 2020c. Teknologiaa työntekijöiden ja asiakkaiden hyvinvointiin – mihin tarpeeseen hyvinvointiteknologiaa? [Technology for the welfare of workers and clients: What needs does welfare technology satisfy? In Finnish; English and Swedish subtitles]. An animation. LUT University. Lahti, Finland. https://www.youtube.com/watch?v=A5Ed7Gdn6PE&feature=youtu.be
Nordic Welfare Centre. 2020. Welfare Technology. https://nordicwelfare.org/en/welfarepolicy/welfare-technology/ (accessed 27 June 2020).
Östlund, B., Olander, E., Jonsson, O. and Frennert, S. 2015. STS-inspired design to meet the challenges of modern aging. Welfare technology as a tool to promote user driven innovations or another way to keep older users hostage? Technological Forecasting and Social Change, 93: 82–90. https://doi.org/10.1016/j.techfore.2014.04.012.
Passini, R. 1999. Information design: An old hag in fashionable clothes. pp. 84–86. In: Jacobson, R. [eds.]. Information Design. The MIT Press. Cambridge, Massachusetts.
Pettersson, R. 1998. Information Design. Mälardalen University. Eskilstuna, Sweden.
Pfannstiel, M.A. and Rasche, C. 2019. Service design and service thinking in healthcare and hospital management. Springer. https://doi.org/10.1007/978-3-030-00749-2.
Pierce, E., Kahn, B. and Melkas, H. 2006. A Comparison of quality issues for data, information, and knowledge. pp. 21–24. In: Khosrow-Pour, M. [ed.]. Emerging Trends and Challenges in Information Technology Management: Proceedings of the 2006 Information Resources Management Association Conference. 17th IRMA International Conference, Washington, DC, USA, May 2006.
Raappana, A. and Melkas, H. 2009. Teknologian hallittu käyttö vanhuspalveluissa: opas teknologiapäätösten ja teknologian käytön tueksi/Helhetssyn vid användning av teknologi inom äldreomsorgen: handbok till stöd för teknologibeslut och användning av teknologi. [Holistic View of Technology Use in Elderly Care Services: Handbook to Support Decisions Concerning Technology and Use of Technology; in Finnish and Swedish]. LUT Lahti School of Innovation. Lahti, Finland.
Roberts, J.P., Fisher, T.R., Trowbridge, M.J. and Bent, C. 2016. A design thinking framework for healthcare management and innovation. Healthcare, 4: 11–14. https://doi.org/10.1016/j.hjdsi.2015.12.002.
Simlinger, P. 2014. What Makes IIID "World Leader in Information Design Development"? International Institute for Information Design. https://www.iiid.net/downloads/IIID-Policy-2014.pdf (accessed 22 June 2020).

Stahl-Timmins, W. 2017. Methods for evaluating information design. pp. 451–462. *In*: Black, A., Luna, P., Lund, O. and Walker, S. [eds.]. Information Design: Research and Practice. Taylor & Francis.

Taneva, I. 2019. Information design. American Economic Journal: Microeconomics, 11(4): 151–185. https://doi.org/10.1257/mic.20170351.

Tufte, E.R. 1990. Envisioning Information. Graphics Press. New Haven, Connec-ticut.

Työppönen, L. 2018. Digitaalisten palveluiden käyttöönotto terveydenhuollossa – tarkistuslistan luominen terveydenhuollon organisaation käyttöön [Implementation of Digital Services in Healthcare: Creating a Checklist for a Healthcare Organization; in Finnish]. Turku University of Applied Sciences. Turku, Finland. https://www.theseus.fi/bitstream/handle/10024/144861/Tyopponen%20Laura.pdf?sequence=1 (accessed 27 July 2019).

Wand, Y. and Wang, R.Y. 1996. Anchoring data quality dimensions in ontological foundations. Communications of the ACM, 39(11): 86–95.

Wang, R.Y. and Strong, D.M. 1996. Beyond accuracy: What data quality means to data consumers. Journal of Management Information Systems, 12(4): 5–34.

Appendix 1. Datasets and analysis

Datasets

Caregivers' focus group interviews concerned implementation-process-related needs. The semi-structured interviews were conducted in two round-the-clock care homes operated by private service providers. Both units are relatively new; one has been operating since September 2018 and the other since May 2019. The first interview took place in March 2019 (unit 1) and the second in May 2019 (unit 2). The interviews were taped. In unit 1, three caregivers aged 20–45 participated. Central themes were the current situation of technology use in the unit, the staff's views concerning technology use in an ideal situation and practices in implementation. In unit 2, six caregivers aged 25–55 participated. Central themes were experiences of new technology use, challenges of new technology use, experiences of good orientation and caregivers' views of clients' experiences of technology implementation. In addition, the unit 2 participants filled in a questionnaire with additional questions on usefulness of technology, clients' experiences and the level of orientation given.

Technology workshops were organized to gain a wider perspective on implementation and orientation to technology. The needs of care managers, caregivers and end-users were examined. Three similar workshops were organized in June 2019 for these groups. An inductive approach was used in planning the workshops. The interviews proceeded inductively, from device-specific questions onwards to orientation-related questions concerning the whole social and healthcare field. There was first a brief introductory lecture on welfare technology. After that, the participants got to know, in a guided way, about four devices: the Joy for All cat, the Gyenno spoon, the Oculus VR glasses and the Laevo exoskeleton. A group interview took place after this. Central themes were orientation needs and prospects of use. The same questions were asked about each device; after that, questions were asked that summarized the whole. Each workshop took about 1.5 hours. The researcher acted as the facilitator. Three other people participated in giving guidance on the devices.

The workshop participants were invited via e-mail. The research group's existing contacts were utilized to recruit them. Three people aged 35–45, and with professional titles of care home manager or director participated in the managers' workshop. Five caregivers aged 35–65 participated in the caregivers' workshop. Their professional titles were either instructor, client instructor or occupational therapist. The end-user workshop had four participants aged 75–90. They had, prior to retirement, worked as a journalist, hospital assistant, storekeeper and truck driver. They all lived in their own apartments in an assisted living facility.

The Gyenno spoon pilot was conducted to determine the perspective of home use. The pilot was implemented in July 2019 with two end-users who both used the spoon for two weeks. Three semi-structured interviews were conducted with both end-users. The first interview concerned the backgrounds of the users and their expectations for use. The intermediary and final interviews concerned thoughts and experiences of use. The interviews were taped. The users also filled in a form to monitor usage. The forms complemented the other data collection by focussing on finding factors that advanced or hindered the use and experiences of orientation that was given by the researcher. The orientation related to purpose of use, structure and functions. It was based on the manual and the researcher's experimentation. Both users received the same orientation. Both users were women of 50–60 years of age and on sickness pension due to their illnesses (Parkinson's disease and dystonia).

Data analysis

All interviews were transcribed. A thematic analysis was then conducted. The survey data were analyzed with Excel. After the thematic analysis concerning needs for implementation and orientation, the Wang and Strong framework of information quality attributes was utilized to distinguish and disaggregate needs for information design, i.e., to find out what attributes should be focussed on, in particular, in information design for each group—older end-users, professional caregivers and care managers.

The research was conducted using ethical standards to avoid any participant harm. The care professionals and clients gave their informed consent to participate in the sessions and research interviews. Safety was ensured by thorough orientation or using the devices only under appropriate, competent control and supervision. The research material was anonymised, and no personal information could be identified from the data.

Appendix 2. The results concerning Wang and Strong's (1996) information quality attributes.

Information quality attribute	Older people	Caregivers	Care managers	Definition
Believability	X			The extent to which information is accepted or regarded as true, real and credible.
Value added		X	X	The extent to which information is beneficial and provides advantages from its use.
Relevancy	X	X	X	The extent to which information is applicable and helpful for the task at hand.
Accuracy				The extent to which information is correct, reliable and certified free of error.
Interpretability	X		X	The extent to which information is in appropriate language and units and the information definitions are clear.
Ease of understanding	X	X	X	The extent to which information is clear without ambiguity and easily comprehended.
Accessibility	X	X		The extent to which information is available or easily and quickly retrievable.
Objectivity	X		X	The extent to which information is unbiased (unprejudiced) and impartial.
Timeliness				The extent to which the age of the information is appropriate for the task at hand.
Completeness		X	X	The extent to which information is of sufficient breadth, depth and scope for the task at hand.
Traceability				The extent to which information is well documented, verifiable and easily attributed to a source.
Reputation	X			The extent to which information is trusted or highly regarded in terms of its source or content.
Consistent representation				The extent to which information is always presented in the same format and is compatible with previous information.

Cost-effectiveness			The extent to which the cost of collecting appropriate information is reasonable.
Ease of operation			The extent to which information is easily managed and manipulated (i.e., updated, moved, aggregated, reproduced, customized).
Variety of information and information sources			The extent to which information is available from several differing information sources.
Concise representation	X		The extent to which information is compactly represented without being overwhelming (i.e., brief in presentation, yet complete and to the point).
Access security			The extent to which access to information can be restricted and hence kept secure.
Appropriate amount of information	X	X	The extent to which the quantity or volume of available information is appropriate.
Flexibility			The extent to which information is expandable, adaptable and easily applied to other needs.

CHAPTER-10

Raising Users' Confidence in their own Technology Literacy as part of the Design Process

Marie Sjölinder

Introduction

With digitalization, interest is growing around the topic of ageing and what it may mean for society to have an increased number of older people in Europe. This has influenced research related to older people and technology (Longhurst et al., 2017; Nikou, 2015; Turkle, 1995). It has also generated a lot of worry related to how older people will manage in the future society, as a high degree of digital literacy will be required (Eriksson and Sjölinder, 2019).

There are many assumptions regarding ageing that are deeply rooted in our culture (DeFalco, 2009). Ageism, a phenomenon that has gained increased attention in recent years, is defined as negative attitudes or behaviours towards individuals based on their age (Lagacé et al., 2015). It is a way of using stereotypes to discriminate against older adults (Butler, 1975), and using prejudice in terms of societal norms that marginalize and make older adults feel unwelcome (McDonough, 2016). These stereotypes rely on a perception of older people as generally unattractive, frail, useless, dependent, forgetful, lacking agency (Lagacé et al., 2015), inflexible and incapable of learning new things (McDonough, 2016). These negative stereotypes also affect social encounters and how people communicate with each other. They affect speech, voice and gestures, and communication is adjusted in such a way as to reinforce and confirm stereotypes (Lagacé et al., 2016).

Research Institute of Sweden, RISE, Email: marie.sjolinder@ri.se

Old-age stereotypes are powerful, and they affect the person that is exposed to them. The stereotypes get integrated into the individual's self-concept and affect the way older adults perceive themselves (Lagacé et al., 2015; McDonough, 2016). This affects the behaviour of the older adults, which in turn confirms the stereotypes (Lagacé et al., 2015). The internalized ageist stereotypes have a detrimental impact on older adults' perceptions of their own competencies (Cherry et al., 2015). They generate feelings of disengagement and contribute to weakening a person's self-esteem and self-efficacy, which is the belief in one's own ability to perform different tasks (Lagacé et al., 2015; Kotter-Grühn and Hess, 2012).

Due to the stereotype that older adults are unwilling and unable to learn how to use new technologies, older adults are often excluded and feel excluded from social issues related to the information society (Lagacé et al., 2015; Mannheim et al., 2019). Age-related presumptions are also used to explain older adults' lower interest and usage of information and communication technologies (ICT). This assumption has been questioned, and the age-related negative stereotypes of being less competent and less interested in learning and using ICT have been discussed in terms of internalized stereotypes resulting in a "self-fulfilling prophecy" (Lagacé et al., 2016), where the older adults actually become less interested in technology since that is what is expected from them. Self-efficacy has also shown to be a determinant of older adults' perception of internet use (Zhang, 2005), and the more seniors endorse ageist stereotypes, the less they use ICT (Lagacé et al., 2015).

The negative stereotypes may further affect the perceived ease of use of the internet due to the assumptions that older adults lack the ability to adapt to new things such as using the internet. If the stereotypes convey the message that the internet is difficult to use for older adults, the threshold for starting to use it becomes higher and the perceived risk of failing becomes larger (McDonough, 2016). As a way of dealing with or protecting self-image due to the internalized age-related stereotypes about the internet and technology use, older adults might underestimate the usefulness of the internet. They may rationalize in terms of that they have managed well without it so far and will continue doing so. This vicious cycle of a negative self-image and an underestimation of usefulness of the internet are likely to further affect attitudes and willingness to engage with new technology (McDonough, 2016).

Being able to categorize things around us and people we meet is necessary for us to be able to handle a complex environment. However, when designing for different user groups, we need to question our own pre-defined views of different user categories. We need to gain an insight into our own prejudices, and make sure that we do not bring them with us into the design of new products and services. Besides trying to avoid being affected by negative stereotypes and prejudices, the design process also provides a possibility to use and apply design methods that increase the participants' belief in their own abilities. This chapter will describe different stereotypes related to older adults and use of digital technology. The chapter also presents how different steps in the design process can be applied and important aspects to consider with the aim of reducing the impact of negative stereotypes and supporting a positive self-image.

Feeling of belonging and belief in one's own ability

Self-esteem is complex and has been discussed from many perspectives in the literature. From a hierarchical perspective, self-esteem has been described in terms of three sub-concepts: *performance* self-esteem, *social* self-esteem, and *physical* self-esteem. The sub-concept relevant in this context is performance self-esteem, which is related to general competence, including intellectual abilities, self-confidence and efficacy (Heatherton and Polivy, 1991). Self-efficacy is related to personal capabilities while self-esteem is the broader concept and related to judgments of self-worth (Bandura, 1997). According to Bandura, people with high self-efficacy believe they can perform well and are more likely to view difficult tasks as something to be mastered rather than something to be avoided (Heatherton and Polivy, 1991; Bandura, 1997).

Self-concept is the view of oneself that is formed through direct experience and evaluations adopted from significant others. Self-concept contributes to an understanding of people's attitudes toward themselves and how they are perceived by others. Theories related to self-concept often address self-image from a holistic perspective, while self-efficacy can vary between different domains (Bandura, 1997).

Self-efficacy has been suggested to play an important role in technology acceptance and use of technology, where negative stereotypes may lead to lowering self-efficacy among older adults, at least if older adults internalize the negative stereotypes and start to believe that they lack the ability to use new technology (McDonough, 2016). There are several studies showing how different aspects related to technology are affected. For example, computer self-efficacy has been shown to be an important predictor of technology usage. Self-efficacy has also been shown to be a predictor of computer anxiety, which has in turn been shown to be related to interest in computers and usage of computers. It can also likely be assumed that the belief in one's own ability, at least to some extent, could explain the age differences in computer interest (Czaja et al., 2006; Ellis and Allaire, 1999).

Much of the human-computer interaction (HCI) research has focused on compensating for age-related decline in terms of different assistive technologies (Rogers et al., 2014). However, older adults today are very different from those of previous generations and there are also large variations between individuals. A large portion of the older population has experience with technology and is used to having access to technical devices and new services. This paves the way for a larger focus on possibilities and benefits with respect to the design of new digital services and how the technology can meet needs in life other than compensating for age-related decline (Rogers et al., 2014), for example in terms of social interaction, self-fulfilment and entertainment. There is a landscape of possibilities, besides creating valuable and meaningful services for older adults, for companies to develop new successful services. It has also been shown that markets are increasingly focusing on the elderly as their primary targets (Pericu, 2017). Both researchers and companies have started to place a larger focus on technology for everyday life—technology to be purchased and used by the older adults themselves. This includes technology and services that older adults are used to having access to in order to be able to manage different tasks or to interact with others. However, to be able to develop successful new services, older adults need to be a part of the design process and share their needs and thoughts about how new technology could be beneficial and useful for them in their everyday lives. A design-thinking approach will not just bring the users and their needs into the design process; it could also be an opportunity to give the participants a joyful experience. Choosing and applying methods for co-creation that enhance the feeling of belonging and being able to contribute will not only increase the value for the participants, but also boost the quality of the outcome in terms of relevant services with functionalities that actually will be used.

Methods for user involvement and design

The following section gives examples of different methods to use and apply when involving users in the design process. Older adults are individuals that differ greatly from one another, with contrasting experiences and lifestyles, and people over 65 years of age today are radically different from people of this age group in previous generations (Pericu, 2017). To gain a better understanding of the specific target group at hand, observations and interviews are methods that can be used in the beginning of the design process to get a deeper understanding of the needs of this group. One further method, which is described below, is developing personas. In this section, the methods are described alongside experiences from applying the method within the context of older adults. At the end of this section, hands-on methods for developing together with the users are presented and elaborated on.

Understanding the user group

Today, technology sometimes plays a greater role in retirees' lives than it did previously during their working lives. Many people who have retired have been using technology both at home and at work for a long time and they have used a variety of services and different devices (Rogers, 2014). From this perspective, it is important to understand the new image of the millennial senior. But to be able to do so, it is also important to understand how other generations perceive them and what expectations are related to old age (Pericu, 2017).

As mentioned, the older adult of today is very different from that of previous generations, and yet the image of an older adult is still very much based on old stereotypes. To be able to approach the field of developing new services useful for older adults, there is a need to go beyond preconceptions and previous attitudes towards what it means to be old. This requires, besides an open mind and self-reflection, a toolkit of methods to gain an understanding of what is meaningful and relevant for different categories of older adults.

Initially, the designers need to learn about the population first-hand. They need to create an exploratory personal view based on a sensitive and critical mind. This step includes, for example, performing observations and conducting interviews (Rebola and Hermann, 2017). This approach sounds easier than it is since, as described above, the population is very diverse.

One way to gain a deeper understanding of the target group at hand is to develop different personas (Cooper, 1999). The aim is to go beyond preconceptions and stereotypes of the target group. The personas should be grounded in empirical work based on gathering information together with the intended target group, or people close to this group if it is not possible to work closely with the intended users. A persona consists of a number of different characteristics describing a typical user in terms of, for example, age, gender, social context, lifestyle and technology usage. They can be used in the design process for reflections about the target group, and the goal of the process is to move from observations to reflective insights (Rebola and Hermann, 2017). The usefulness of using personas was shown in a study by Sjölinder and Scandurra (2015). In this study, a new communication device was developed together with the personnel and the older-adult residents of a nursing home. In the beginning of the project, both the older adults and the personnel were asked about the technology experience of the older adults. The answers given by the older adults themselves were compared with those given by the personnel. The care personnel underestimated the older adults' technology experience and overlooked the fact that many of the older adults had once worked with technology in different environments. When a few months of the project had passed, the personnel participated in a task aimed at forming descriptions of different subgroups of users. The personnel created a number of personas (Cooper, 1999) that represented older adults living at the nursing home. This turned out to be an eye-opener for the personnel since it forced them to reflect on their residents and their relationships to technology. Thinking of a person as he or she used to be was useful, as was reflecting upon the entire person beyond their medical conditions. The task contributed to gaining a new perspective of the elderly since the care personnel were forced to think about them as possible users of new technology. This also increased the personnel's understanding of the possibilities surrounding the technology, and they started to suggest new ways of using the technology at the nursing home (Sjölinder and Scandurra, 2015). In this case, using personas turned out to be effective. However, they need to be developed and used with caution since there is also a risk of manifesting stereotypes and prejudges about a certain group.

Workshops and hands-on experience

Different fields of human-computer interaction (HCI) research originated in the 1980s and the 1990s and have since come a long way towards understanding the importance of involving users in the process of developing new technology. Within the participatory design (PD) framework (Greenbaum, 1991; Schuler, 1993), the degree of user participation may vary, but regardless of activation degree, in PD, developers and practitioners/users are seen as actively cooperating partners. They work together to reduce uncertainty and risk in the development of innovations, where a detailed conception of exactly which future needs should be supported is often lacking (Greenbaum, 1991; Schuler, 1993). Previous research presents several methods for engaging users with the aim to create future environments, e.g., so-called 'future workshops' (Jungk and Müllert, 1987). Other methods to bring analysis of future needs into system development include iterative prototyping and scenario-based design, preferably applied together with potential users in a collaborative approach (Scandurra et al., 2008).

The concept of living labs (Niitamo et al., 2006) takes a more hands-on approach, with the objective of enabling older people to be more directly involved (Rogers et al., 2014). A living lab can be a place where different user groups are invited to learn about and try new technology. The context and the methods used can vary. One way could be to let the participants interact with a physical model or mock-up that can provide more concrete support in envisioning a proposed system (Rogers et al., 2014; Tröger et al., 2016). Less-structured ways of getting in touch with new technology can include arranging informal meetings and technology cafés (Tröger et al., 2017), where possible technology users can meet, learn more and get engaged.

The design-thinking approach paves the way for focusing on user needs and provides support for solving problems in different contexts. Rebola and Herrman (2017) proposed a design-thinking approach targeting an older user group. They suggested a system for design thinking that addressed the needs of older adults using seven phases: position, purpose, prosthetics, place, participation, potential and presentation. In their participation phase, they placed an emphasis on methods and tools for empowering older adults to co-create solutions to their concerns and needs. In this phase, someone familiar with creative problem-solving processes needs to develop methods that allow the older adult to think like a designer and co-create a way to address their concerns. The goal of this step is to develop methods for giving the older adult tools they can use to become a creative partner in addressing problems they have pinpointed. Rebola and Herrman (2017) state that the goal of developing new technology is to provide the ageing populating with creative, attractive and pervasive solutions that avoid physical, visual or experiential segregation. The above relates to promoting the feeling of being capable, instead of creating a feeling of not being able to participate or to contribute.

Rogers et al. (2014) suggested an alternative approach based on empowering people, in this case older adults, by providing them with new technologies and/or teaching them a new skill. In their study, they employed a technical toolkit that the participants could use to solve different tasks. During this process, when playing around together, the participants started to talk both about problems they had with technology and problems they thought could be solved by technology. In that sense, the method was used as a way of supporting the ideation process. Finally, and supported by the work of Rogers et al. (2014), the important aspect of having fun should not be forgotten. This is an important aspect, both to engage people in the design of new technology and with respect to using it. The work conducted by Rogers et al. (2014) supports the importance of getting the feeling of being capable and able to contribute. In the hands-on usage of a technical toolkit, people learn by doing. This contributes to empowerment and creativity, and makes it easier to generate expressive creative output as well as input. By solving problems and mastering the technology, the individuals perceive both the feeling of managing something in the moment, as well as an increased long-term feeling of being someone that can solve technical challenges. However, tools like this need to be easy to use, fun to play with, and empower people to learn and create (Rogers et al., 2014).

A further added value of being part of the design process could be the possibility to be part of developing new technology for the future. The older adults that participate could get access to devices and services that their children and grandchildren are not aware of. The technology could be something to show to the rest of the family, or it could be a family activity to play with together. In the work conducted by Rogers et al. (2014), the older adults wanted to use the technical toolkit together with their children and grandchildren and they thought it could be a way to interact with them on their level and also to some extent impress them with technology they had mastered but their grandchildren had never heard of.

The key is technology and services that there is a need for and that people in the long run find attractive enough to pay for. Another important aspect, as mentioned above, is user involvement and co-creation. However, the importance of the involvement of all stakeholders has also been placed in focus alongside methods that support collaboration between designers, different user groups, developers, managers and researchers (Sjölinder et al., 2016; Rebola and Herrman, 2017). Kopec et al. (2018) place a focus on the actual real-world context of smaller companies with limited capabilities and resources. There has to be a realistic way for the companies to actually develop the ideas emerging from ideation sessions (Sjölinder et al., 2016; Kopec et al., 2018). Workshops involving multiple stakeholders, such as older adults, care personnel, researchers, developers and representatives from the companies, have been shown to increase the feasibility for the innovation companies (Sjölinder et al., 2016). When collaboration between different stakeholders focuses on balancing the demands of the users and the prerequisites of the industry there is a greater likelihood of developing a successful end product. Involving ageing users in finding the balance between user needs and feasibility in the development, as well as active involvement from industrial partners, may contribute to a common understanding in developing realistic and meaningful services. For example, Kopec et al. (2018) suggested a structured way of engaging older adults in the agile development. Their SPIRAL (Support for Participant Involvement in Rapid and Agile software development Labs) method has the goal of adding sustainability and flexibility to the development process. The method has a participatory approach and is based on a living lab concept. The method consists of strategies for involving older adults in the software development processes to support the agile approach with rapid prototyping. Finally, another important aspect for the companies is to find a market that is large enough to motivate the costs for development. Often there is a trade-off between addressing user needs of specific groups and developing technical solutions that can be offered to a larger market (Sjölinder et al., 2016).

Aspects to consider during the design process

When designing for and together with older adults (and other user groups), it is not just which methods to use that need to be considered; several aspects surrounding the design sessions are crucial. The social context is important, both for older adults and for the personnel that participate. The participants need to feel comfortable and included, and the methods used have to be adapted to the specific context. Another important aspect to consider is the size of the group and the background of the participants. Decisions about how to put the group together need to be taken based on goal and purpose, and on the expected outcome of the design session. In the section below, some important practical and social aspects to keep in mind when planning and arranging the participatory design sessions are presented.

Arranging focus groups and workshops

A set of guidelines on how to arrange workshops and focus groups was suggested by Mannheim et al. (2019). They propose a holistic view of competence, a view which goes beyond cognitive ability and which takes sensory decline into consideration. The guidelines provide advice regarding ethical aspects and considerations. Mannheim et al. (2019) suggest awareness of stereotypes and use of a universal design so as to not exclude older adults from being able to use the technology. Further, the guidelines present recommendations with respect to context and awareness of the participant's situation; for example, to choose a time and place that are convenient or to try to use their natural environment. The guidelines also place an emphasis on making it easy to participate and to withdraw if so needed. All material used in the design process should be easy to read and easy to use to avoid the feeling of not being able to participate, and all aspects and consequences related to privacy and confidentiality should be clearly stated.

For the development of new successful services, it is, as often stated, important to have user involvement and feedback from users. To get both older adults and others committed to this, the participation has to provide some kind of added value. If the development is conducted in an iterative way with continuous feedback from the user group, feedback must also be given back to the user group making them understand that their contribution is valuable. Important aspects of user engagement are moving from designing for frailty towards focusing on empowerment and being able to contribute and have the opportunity to learn new things (Rogers et al., 2014).

Creating a social context

Social belonging and being part of a group is an important aspect of our lives. Being part of a design process could contribute to fulfilling this need (Sjölinder et al., 2016). Being part of the design process may, for the participants, increase the feeling of belonging to a social context or to a group. Besides making a valuable input to the design process, arranging design workshops with older adults could provide an important and meaningful social context. The possibility to meet on a frequent basis and make new friends could both increase social inclusion and contribute to engaged workshop participants (Sjölinder et al., 2016). Being part of a group and feeling that we contribute in a meaningful way are important aspects that should not be forgotten.

Co-design could be conducted in a way in which the participants get a positive experience, feel like partners and engage in the design process between the sessions (Mannheim et al., 2019). Creating the possibility for the same participants to meet on a frequent basis gives them the chance to get to know each other and plan activities outside the focus groups. The participation in this social context and in the involvement of developing new technology could also be something to be proud of and talk about with friends and family. The importance of being seen and appreciated by others is also crucial to consider when developing services targeted towards older adults (Sjölinder and Scandurra, 2015). Workshops should also be playful and collaborative to inspire creativity and inventiveness, and they should include activities to provide training followed by free time to play and discuss new ideas (Rogers et al., 2014).

The social context and possibility for the personnel to understand and convey the needs of older adults are also important and should be part of the design process. It is crucial that the personnel are part of the development and that they have enough time to understand the needs of the older adults. Often people are not explicitly aware of their problems and needs themselves and cannot explain or answer questions about them. It can therefore be a challenge to gain insight about the needs of users or the problems they are facing (Rogers, 2014). When the personnel are given the time and the possibility to actually get involved in the design process and the needs of the older adults, their view of what actually is needed among the older adults could change. This contribution to the design process could be very valuable since in this situation, the personnel can suggest new ways of using the technology that were not thought of before (Sjölinder and Scandurra, 2015).

Mixing people with different backgrounds

Communication and working together in cross-generational groups may increase understanding and reduce negative stereotypes (Lagace et al., 2016). In a study Pericu (2017) let two different generations, seniors and millennials, use their contrasting experiences and their points of convergence to collaborate on a design process. The synergy between the young designers and the older-adult users became an opportunity to engage in people's needs and to bring different points of view into the design process. Collaborations between design students and older adults could also become a way to broaden perspectives and make the young designers view the ageing process and ageing differently (Coleman and Myerson, 2001). Cross-generational or cross-group workshops can be applied in several ways and the advantages can be many. Both older adults and care personnel could be users of different technologies, but with different goals. The care personnel might use the technology as part of their work, and therefore their needs and preferences are related to achieving goals related to their working context. The older adults might use the technology either as part of their care or for other everyday life tasks. Cross-group workshops between older users and care personnel could increase the understanding that the groups have different goals for using the technology, and it could also contribute to finding common goals and ways of using the technology that will be beneficial for all user groups. These cross-group sessions can also bridge the gap between the older adults and the personnel with respect to understanding each other's needs for technology, and can contribute to a deeper understanding among the personnel that the older adults actually could use and benefit from new technology (Sjölinder et al., 2018).

Furthermore, it could be beneficial to include other relevant groups from the very beginning in the design process. Specialists in different areas could contribute in order to help understand specific parts of the problem. For example, in a study conducted by Pericu (2017), geriatricians were involved in the design process of a service for daily life activities to support a healthier lifestyle. In this case, they contributed with knowledge about medical information, health priorities and prevention measures related to seniors. This increased the knowledge in the team about main diseases of the elderly population and increased the understanding for common situations related to ageing.

We know from HCI research that usability aspects should be brought in early in the development process (Lockwood, 1999), so another important group to involve is the developers. By including developers in the meetings from the very beginning, they get a first-hand understanding of the problems different user groups are facing, something that is very difficult to get in other ways without being present in the discussions. A key to success is the responsiveness of the developers, which can only be reached by true engagement and by participation in settings where all stakeholders are encouraged and guided to perform at their best. One example of a method that has worked well is observation of real usage performed by the entire team, including the developers. User profiles or personas could also be useful for the developers' understanding of the environment and the users that the services should support (Sjölinder et al., 2016).

Testing the technology together

Sharing the experience and being able to play together could have several advantages. It could be a catalyst for imagining, free thinking and exploring, and could create a positive and relaxed atmosphere in which people could freely bounce ideas around (Rogers et al., 2014). In a study by Rogers et al. (2014), a technical toolkit was used in a series of workshops in which the participants both had fun together and learned new things. Another way to create this sort of atmosphere could be to set up sessions in which older adults and personnel can test new technology together. This could provide insights from several perspectives and could also be a nice and fun experience for the participants (Sjölinder et al., 2014). After testing the technology, older adults and personnel could be given similar questions to answer. The answers provided by the older adults could give hands-on feedback about usage, and the answers given by the personnel could provide deeper insights into what could be difficult to use and about which features to include to a greater extent (Sjölinder et al., 2017). A shared experience between care professionals and older adults could also contribute to a cross-generational common ground for communication and understanding with respect to technology usage (Sjölinder and Scandurra, 2015; Sjölinder et al., 2017).

Introducing new technology

According to findings showing that beliefs about one's own ability are related to how we approach new technology, the manner in which technology is presented and introduced is crucial. It is also important how the individual is treated as a possible user of the technology. As described previously, there is a risk of decreased engagement and interest when older adults are met with age-related stereotypes. However, it has been shown that age differences related to technology usage can be reduced depending on how the older adults are treated or involved (Broady et al., 2010).

Sense of identity and belonging in a context are important. This includes a personal sense of being part of a certain context and also the feeling of being accepted as an individual with unique needs for personal space and companionship (Goonetilleke and Karwowski, 2019). Previous work has pointed out the importance of avoiding stigmatizing the external appearance of products (Eriksson and Sjölinder, 2019; Mannheim et al., 2019), since products communicate and spread a message (Pericu, 2017), for example, through the design of products and services, and the instructions that belong to the devices or services (Eriksson and Sjölinder, 2019). When designing new technology, stereotypes of frail older adults often affect designers and might shape how the technology targeted towards older adults is designed (Mannheim et al., 2019). From a psychological perspective, this relates to concepts of self-image, social interaction with others, and how we are perceived by others. From a marketing perspective, it is also a disadvantage if products are designed and recognized as targeted towards older adults, as these products will be rejected by their target audience. The design focus should instead be on a multigenerational customer (Pericu, 2017). Mannheim et al. (2019) suggest the adoption of a universal design, which enhances the feeling of inclusion and being part of a group or a social context.

The ideation and the co-creation process could also contribute to technology adoption and to the domestication process, and help increase interest in technology and willingness to engage in, and use, new technology. This was shown and explained by Valk et al. (2017), in terms of defining the meaning and value of the technology during the ideation process for the end users. One example of where the co-creation process contributes to the domestication process is the use of technology ambassadors among the personnel. These ambassadors can spread engagement, among both personnel and older adults, and they can gain insights about needs and possible further solutions (Sjölinder et al., 2018).

Supporting the learning process

It is often claimed that it is more difficult for older adults to learn how to use technology. Studies have shown that the learning process is longer (Kelly and Charness, 1995), and that more time is needed to solve different tasks (Kubeck et al., 1999; Mead et al., 2000; Sjölinder et al., 2003). However, the effect of these age-related differences might be overestimated. Studies reporting difficulties for older adults who are learning and using technology have often failed to provide the appropriate context for introduction and learning (Broady et al., 2010). This again reiterates the importance of not contributing to a degrading self-image but instead providing a supportive environment (Broady et al., 2010).

The possibility to teach others has positive effects, and ambassadors or lead users, both among older adults and among the personnel, have shown to be beneficial (Mannheim et al., 2019; Sjölinder and Scandurra, 2015). These roles convey both a feeling of having knowledge and a feeling of being able to contribute in a meaningful way. From a personnel perspective, it provides the possibility to take the lead in issues related to technology and to be someone who has knowledge and skills to share with others. These personnel can involve older adults and relatives and create a positive and engaging atmosphere surrounding technology, where new ideas for technology usage can be brought to the surface (Sjölinder and Scandurra, 2015). With respect to older adults, they can be involved in the design process in terms of being a part of leading the focus groups; for example, by engaging retired designers in leading parts of the design process, or by teaching the methods for turning problems into creative solutions (Rebola and Herrman, 2017). Mannheim et al. (2019) also showed that networks supported by technology-literate mediators, for example older adults, could create digital support circles and act as agents for enhancing wider inclusion. There are of course a number of other ways of using both older adults and personnel as technology ambassadors. However, the important aspect is that they create a social context and that they provide the possibility to learn new skills that can be conveyed to others.

Discussion and conclusions

This chapter has described different stereotypes related to older adults and the design and use of digital technology, and also how these stereotypes affect older adults' views of themselves and of their technology usage. The negative age-related stereotypes could contribute to feelings of reduced social belonging, both with respect to society in general, and in regard to technology usage. The negative stereotypes could further contribute to reduced confidence in one's own ability and in feelings of being less able to contribute. If designers, researchers and developers are not aware of the age-related stereotypes when they involve older adults in the design of new technology, there is a risk that these stereotypes are strengthened and that the older participants internalize the negative stereotypes even further. However, awareness and avoidance of stereotypes and prejudices will provide a good starting point. When engaging in the various phases of the design process, it is possible to actively apply methods in ways that support the feeling of social belonging, the feeling of being capable and being able to contribute.

In the initial phase of understanding the users and their problems, it is important that all stakeholders are aware of existing age-related stereotypes and try to avoid them instead of reinforcing them. There has to be an understanding that older adults today have very little in common with previous generations of older adults, and that there are large variations between individuals. One way of gaining a better understanding of the older adults' relationships to technology is to develop different user profiles or personas. This could contribute to a better understanding of the older adults as technology users. Cross-generational groups could also increase understanding and broaden different perspectives. Using groups consisting of different stakeholders and people with different roles from the very beginning increases the possibility of gaining a better understanding of the problem. From the perspective of the older adult, it could contribute to a feeling of being part of a group and to the feeling of being able to contribute.

Different methods in the design process provide many ways of involving older adults in a way that could add value for the participants. Living labs and technology cafés are setups that support social interaction and being part of a group. People can get together, meet and discuss new technology. In these contexts, the participants have the possibility to find out about and try new technology. By doing so, they have the possibility to increase their feeling of having knowledge, both in terms of awareness of technology on the frontline and in terms of managing it. The participation and this knowledge could feed into a self-image of someone who is familiar with new devices and services, and it could serve as a topic for discussion with friends and family. Hands-on methods for ideation, such as using different technological toolkits, could engage participants and provide insight in terms of which activities to conduct. When conducted in a way that does not exclude anyone, the activity could be something that is fun and interesting and could help play down challenges related to technology usage. However, it is important that the atmosphere is open and friendly, that the tasks are manageable, and that the participants can solve these tasks together without pressure (Rogers et al., 2014). The atmosphere and the possibility to have fun also contributes to more open discussion in which problems, needs and ideas for new solutions can more easily be expressed and addressed. Besides these social advantages of getting to know new people and having fun, by solving tasks using technology, the participants have the opportunity to leave the sessions with new knowledge and a feeling of being able to manage new technology. The fun part and the feeling of having a nice time cannot be emphasized enough. Many brainstorming methods consist of procedures that could have several disadvantages, or at least aspects that need to be carefully considered. The procedures and the different steps might not fit everyone, and people could be uncomfortable with expressing their ideas or presenting outcomes of discussions in front of a group, especially if it has been difficult to participate actively due to a loss in hearing and/or vision (Mannheim et al., 2019). Methods and materials therefore have to be developed with this in mind, and also with a focus on making it fun without being childish. If this is not done in a very thoughtful and careful manner, the participants could leave with a decreased belief in their own ability to perform and to contribute in different contexts.

The design process also provides many options to involve future users in ways that create added value for the participants. Being part of a social context where it is also possible to get new insights about technology could in itself be perceived as a positive experience. However, it is important that the context is set up in a way where it does not create feelings of not being able to deliver the output that is asked for. Everything around the setup needs to be considered, from information material to technical support. If the participants try or use the technology in their homes or outside the focus group context, a thorough plan must be in place for dealing with problems they might encounter. There must be a clear message that the participants' feedback and reporting of problems is valuable, without implying that they have difficulties handling new technology. Regardless of what issues they face, things could be improved regarding the technology, the instructions, or the setting around the usage or the test. A further way to add value for the participants is to let the same people meet or participate on a frequent basis. This has advantages both in terms of possible social interaction, and in terms of quality of the feedback. Meeting the same people, from a project or a company, several times strengthens the bonds and makes it easier to report errors and provide negative feedback about the product or the service. For the participants, it will create a stronger feeling of being able to contribute. Finally, by developing services targeted at a broad audience and by involving participants from different age groups during development, stigmatization related to technology for older adults could be reduced.

Development and implementation of new technology could benefit from the use of technology ambassadors. The feeling of being selected, and the opportunity to learn more and to teach others, contributes to the feeling of being capable. When provided with enough time and support, the technology ambassadors could become really engaged and not only spread the technology to others, but also gain a better understanding of which needs could be met by new technology and new digital services. The technology ambassadors could furthermore play an important role when implementing new technology in an organization, as they will be engaged users that support others and spread a positive atmosphere surrounding the usage of the technology. Finally, by involving the ambassadors (and other future users) from the very beginning during the ideation phase, the co-creation process in itself could contribute to technology adoption (Valk et al., 2017) because it will place an emphasis on the meaningfulness and value of the technology.

References

Bandura, A. 1997. Self-efficacy: The Exercise of Control. W H Freeman/Times Books/Henry Holt & Co. New York, USA.

Broady, T., Chan, A. and Caputi, P. 2010. Comparison of older and younger adults' attitudes towards and abilities with computers: implications for training and learning. Br. J. Edu. Technol., 41(3): 473–485. doi:10.1111/j.1467-8535.2008.00914.x.

Butler, R.N. 1975. Why Survive? Being Old in America. Harper & Row, New York, USA.

Cherry, K.E., Allen, P.D., Denver, J.Y. and Holland, K.R. 2015. Contributions of social desirability to self-reported ageism. J. Appl. Gerontol., 34: 712–733.

Coleman, R. and Myerson, J. 2011. Improving life quality by countering design exclusion. Gerontechnology, 1(2): 88–102.

Constantine, L. and Lockwood, L. 1999. Software for Use: A Practical Guide to the Essential Models and Methods of Usage-Centered Design. Addison-Wesley, Reading, MA, USA.

Cooper, A. 1999. The Inmates Are Running the Asylum. Macmillan Publishing Co., Inc. Indianapolis, IN, USA.

Czaja, S.J., Charness, N., Fisk, A.D., Hertzog, C., Nair, S.N. et al. 2006. Factors predicting the use of technology: Findings from the Center for Research and Education on Aging and Technology Enhancement (CREATE). Psychol. Aging., 21: 333–352.

DeFalco, A. 2009. Uncanny Subjects. The Ohio State University Press, Columbus.

Ellis, D. and Allaire, J.C. 1999. Modeling computer interest in older adults: The role of age, education, computer knowledge, and computer anxiety. Hum. Factors, 41: 345–355.

Eriksson, Y. and Sjölinder, M. 2019. The role of designers in the development and communication of new technology. pp. 37–48. In: Sayago, S. Sergio [ed.]. Perspectives on Human-Computer Interaction Research with Older People. Springer International Publishing. doi:10.1007/978-3-030-06076-3.

Goonetilleke, R.S. and Karwowski, W. 2019. Enhancing the Life of the Elderly—An Application of Design Thinking AHFE 2019, AISC 967, pp. 388–396, 2020. Springer Nature Switzerland AG 2020.

Greenbaum, J. and Kyng, M. 1991. Design at Work: Cooperative Design of Computer Systems. Lawrence Erlbaum Associates, New Jersey, USA. pp. 3–24.

Heatherton, T.F. and Polivy, J. 1991. Development and validation of a scale for measuring state self-esteem. J. Pers. Soc. Psychol., 60: 895–910.

Jungk, R. and Müllert, N. 1987. Future Workshops: How to Create Desirable Futures. Institute for Social Interventions, London, UK.

Kelly, C.L. and Charness, N. 1995. Issues in training older adults to use computers. Behav. Inf. Technol., 14(2): 107–120.

Kopec, W., Nielek, R. and Wierzbicki, A. 2018. Guidelines towards better participation of older adults in software development processes using a new SPIRAL method and participatory approach. pp. 49–56. In: Proc. of CHASE'18: IEEE/ACM 11th International Workshop on Cooperative and Human Aspects of Software, Gothenburg, Sweden. CHASE'18.

Kotter-Grühn, D. and Hess, T.M. 2012. The impact of age stereotypes on self-perceptions of aging across the adult lifespan. The Journals of Gerontology: Series B, September 2012, 67(5): 563–571, doi:10.1093/geronb/gbr153.

Kubeck, J.E., Miller-Albrecht, S.A. and Murphy, M.D. 1999. Finding information on the World Wide Web: Exploring older adult's exploration. Educ. Gerontechnology, 25: 167–183.

Lindsay, S., Jackson, D., Schofield, G. and Olivier, P. 2012. Engaging older people using participatory design. In Proc. CHI 2012, ACM Press, pp. 1199–1208.

Longhurst, B., Smithe, G. and Bagnall, G. 2017. Introducing Cultural Studies. Routledge, London.

Lu, Y., Valk, C.A.L., Steenbakkers, J.J.H., Bekker, M.M., Visser, T. et al. 2017. Can technology adoption for older adults be co-created? Gerontechnology, 16(3): 151–159. doi: 10.4017/gt.2017.16.3.004.00.

Mannheim, I., Schwartz, E., Xi, W., Buttigieg, S.C., McDonnell-Naughton, M. et al. 2019. Inclusion of older adults in the research and design of digital technology. Int. J. Environ. Res. Public Health, 16(19): 3718; doi:10.3390/ijerph16193718.

McDonough, C.C. 2016. The effect of ageism on the digital divide among older adults. J. Gerontol. Geriatr. Med., 2(1): 1–7. doi:10.24966/GGM-8662/100008.

Mead, S.E., Sit, R.A. and Rogers, W.A. 2000. Influences of general computer experience and age on library database search performance. Behav. Inf. Technol., 19(2): 107–123.

Nikou, S. 2015. Mobile technology and forgotten consumers: The young elderly. Int. J. Consum. Stud., 39(14): 294–304

Niitamo, V.-P., Kulkki, S., Eriksson, M. and Hribernik, K.A. 2006. State-of-the-art and good practice in the field of living labs. In Technology Management Conference (ICE), 2006 IEEE International. IEEE, p. 1–8.

Pericu, S. 2017. Designing for an ageing society: products and services. Des. J., 20: sup1, S2178-S2189, doi: 10.1080/14606925.2017.1352734.

Rebola, C.B. and Hermann, E. 2017. Design Thinking as a Process for Innovative Older Adult Applications. ACHI 2017: The Tenth International Conference on Advances in Computer-Human Interactions, March 19–23, 2017, Nice, France.

Rogers, Y., Paay, J., Brereton, M., Vaisutis, K., Marsden, G. et al. 2014. Never Too Old: Engaging Retired People Inventing the Future with MaKey. CHI 2014, April 26–May 1, 2014, Toronto, ON, Canada. doi:10.1145/2556288.2557184.

Scandurra, I., Hägglund, M. and Koch, S. 2008. From user needs to system specifications: Multi-disciplinary thematic seminars as a collaborative design method for development of health information systems. J. Biomed. Inform., 41(4): 557–569.

Schuler, D. and Namioka, A. [eds.]. 1993. Participatory Design Principles and Practices. Lawrence Erlbaum Associates Inc., London, UK.

Sjölinder, M., Höök, K. and Nilsson, L.-G. 2003. The effect of age-related cognitive differences, task complexity and prior internet experience in the use of an on-line grocery shop. Spat. Cogn. Comput., 3(1): 61–84.

Sjölinder, M. and Scandurra, I. 2015. Effects of using care professionals in the development of social technology for elderly. pp. 181–192. In: Zhou, J. and Salvendy, G. [eds.]. Human Aspects of IT for the Aged Population. Design for Everyday Life. ITAP 2015. Lecture Notes in Computer Science, vol. 9194. Springer, Cham. doi:10.1007/978-3-319-20913-5_17.

Sjölinder, M., Scandurra, I., Avatare Nöu, A. and Kolkowska, E. 2016. To meet the needs of aging users and the prerequisites of innovators in the design process. pp. 92–104. In: Zhou, J. and Salvendy, G. [eds.]. Human Aspects of IT for the Aged Population. Design for Aging. ITAP 2016. Lecture Notes in Computer Science, vol. 9754. Springer, Cham. doi:10.1007/978-3-319-39943-0_10.

Sjölinder, M., Scandurra, I., Avatare Nou, A. and Kolkowska, E. 2017. Using care professionals as proxies in the design process of welfare technology—perspectives from municipality care. pp. 184–198. In: Zhou, J. and Salvendy, G. [eds.]. Human Aspects of IT for the Aged Population. Aging, Design and User Experience. ITAP 2017. Lecture Notes in Computer Science, vol. 10297. Springer, Cham.

Sjölinder, M., Avatare Nöu, A. and Fristedt, J. 2018. ICT services for nursing homes—A needs analysis. The 11th World Conference of Gerontechnology, May 7–11, 2018, St. Petersburg, Florida, USA.

Tröger, J., Alexandersson, J., Britz, J., Rekrut, M., Bieber, D. et al. 2016. Board games and regulars' tables—Extending user centred design in the Mobia Project. pp. 129–140. In: Zhou, J. and Salvendy, G. [eds.]. Human Aspects of IT for the Aged Population. Design for Aging. ITAP 2016. Lecture Notes in Computer Science, vol. 9754. Springer, Cham. doi:10.1007/978-3-319-39943-0_13.

Tröger, J., Mariano, J., Marques, S., Mendonça, J., Girenko, A. et al. 2017. Technology experience café—Enabling technology-driven social innovation for an ageing society. pp. 199–210. In: Zhou, J. and Salvendy, G. [eds.]. International Conference on Human Aspects of IT for the Aged Population. Springer, Cham.

Turkle, S. 1995. Life on Screen. Identity in the Age of Internet. Simson & Schuster, New York, USA.

Zhang, Y. 2005. Age, gender, and Internet attitudes among employees in the business world. Comput. Hum. Behav. 21: 1–10. doi: 10.1016/j.chb.2004.02.006.

CHAPTER-11

Lessons Learned
A Plea for Curricularizing Design Thinking in Engineering Education

Anne Wallisch[1,*] and *Kristin Paetzold*[2]

Introduction

Design thinking has its origins in an extracurricular course for mechanical engineers. These roots are taken up here and design thinking is used as a didactic element to make content of the lecture 'Methods of Product Development', aimed at students of mechanical and aerospace engineering, tangible. In this chapter, we reveal the potential of integrating such a course into the regular standard curriculum by reporting the results achieved and the experiences made within our pilot course. By assessing the design thinking experience of engineering students and older people invited to the classroom and participating as representatives of the targeted user group, we see considerable potential in using design thinking as strategy to sharpen the engineering novices' view towards humans in their roles as users. Sensitizing newbies for the need to not only observe typical actions, live situations and daily routines of the targeted users but also to understand the appearing challenges from within these users' perspective, fosters their capability of deriving requirements more precisely. From an organisational perspective, students used design thinking to change processes and combined activities to improve user satisfaction, such as better identification of needs by approaching user perspectives more holistically. From the students' point of view, design thinking fostered their reflecting on what customers really want and how to create appropriate processes, actually making product innovations

---

[1] Universität der Bundeswehr München, Werner-Heisenberg-Weg 39, 85577 München, Germany.
[2] Technische Universität Dresden, Institut für Maschinenelemente und Maschinenkonstruktion, Professur für Virtuelle Produktentwicklung, 01062 Dresden, Email: kristin.paetzold@unibw.de
* Corresponding author: anne.wallisch@unibw.de

possible. As our world is rapidly changing, tomorrow's engineers must be prepared to (co-)designing change. This chapter illustrates a promising way of its accomplishment.[1]

Back to the roots

Designing everyday objects is not always intuitive and even if so; sometimes designers have something different in mind as their users do. In return, sometimes users feel disappointed by a product even if, from the design point of view, designers did a very great job. To tackle complex issues, designers created generations of design approaches since the late 1960s and most of its techniques were generated by various sources throughout the decades since then. In the early 1970s, for example, Horst Rittel and Melvin Webber (1973) debated the idea that problems could be quantified, understood and could have a clear path of action and suggested that designers "ought to be asking what systems do, instead of what they are made of, and more so who they are built for".[2] In line with that, in the 1980s, Nigel Cross (1982) defined "practicality", "ingenuity", "empathy", and "a concern for 'appropriateness'" as being the values of design.[3] At the same time, Donald Norman and Stephen Draper (1986) originated the term user-centred design to describe design processes, in which users' influence how a design takes shape.[4] Norman (1988) recognizes the needs and interests of users and recommends placing them at the centre of the design. However, human beings are extremely diverse, individualized and highly contextualized, which makes it very challenging to generalize their opinions and preferences. Unsurprisingly, a number of product failures testify that engineering designer's assumptions on what particular user groups may like or not like are often quite wrong.

To overcome this obstacle, starting in Scandinavia, the ideas of participatory design, where designers act as curators of experiences that involve the whole community, gained momentum (Sanders and Stappers,

---

[1] To render the chapter easy to read, it addresses groups of persons (e.g., older people, students, designers) in a neutral form, always referring to both male and female persons.

[2] Horst Willhelm J. Rittel (1930–1990) was a design theorist and university professor. His field of work is the science of design, or, as it also known, the area of design theories and methods (DTM), with the understanding that activities like planning, engineering, policy making are included as particular forms of design. Melvin M. Webber (1920–2006) was Professor of City Planning and Chairman of the Center for Planning and Development Research at the University of California, Berkeley.

[3] Nigel Cross (born 1942) is a British academic, a design researcher and educator, emeritus professor of Design Studies at The Open University, UK, where he was responsible for developing the first distance-learning courses in design.

[4] Donald A. Norman (born 1935) is an American researcher, professor, and the director of The Design Lab at University of California, San Diego, where Stephen W. Draper (born 1950), a British researcher, worked in the Cognitive Science laboratory of the Psychology department, and at Glasgow University from 1987.

2008). The growth design experienced since then was enhanced by the perception of designers as being kind of a medium through which innovation was triggered (Brown, 2009), while non-designers were introduced to tools and techniques developed to enhance participation (Sanders and Stappers, 2014). Nevertheless, user participation processes are dynamic and complex, and therefore more challenging than closed innovation processes. They place new demands on the development process, in particular the need for a new, open mindset and approaches to support designers in addressing both the user's demands and their expertise in the solution finding (Goodman-Deane et al., 2008; Wallisch and Paetzold, 2020). Surprisingly contrasting, these aspects only recently found their way into engineering education, organized as extracurricular activities mostly, which we assume as being the reason for the fuzzy user-centred design practice observed (Wallisch et al., 2019). Many activity examples of so-called user interaction turned out limited to testing existing solutions sporadically with customers or other stakeholders to verify the previously supposed user needs (Wallisch and Paetzold, 2018). Engineers and designers are working on future ideas and how to improve complicated existing products to more user-approachable ones, however, without having continuous interaction with their users (Carlgren et al., 2016). Organizations often seem to lack knowledge and resources to conduct changes fully and strengthen the role of involving users in existing workflows. Moreover, even if they conduct testing, this mostly happens either remotely with limited coincidental access to users or in-house only with access to people developing products and focusing on user interfaces instead of user experiences (Kosmala et al., 2019).

Therefore, on the one hand, we must recognize a general lack of knowledge and skills on implementing user-centred thinking of design. On the other hand, simultaneously, design thinking as a concept to approach user-centred design has become increasingly popular across disciplines (Bouwman et al., 2019). With some see it as a mere toolbox of methods, while others see it as an umbrella term for a mindset that determines how designers think and act, it has become a somewhat ambiguous term (Kimbell, 2011). As said before, the traditional interpretation of design thinking in research focussed on fundamental cognitive acts of designing, such as information search and generation, mental imaginary, assessment and evaluation, structuring and learning (Bayazit, 2004; Goldschmidt and Badke-Schaub, 2009; Huppatz, 2015). The current conception of design thinking has a prominent role as established corporations ramp up innovation efforts focusing on user needs.

The plurality of appreciations, conceptualizations and names of design thinking reflect lacking curricular standards in design thinking in academic engineering education. Instead of teaching thinking design from the user's perspective, simply because it is users who designs are for, students are educated to think in technical solutions prior to awareness of the user's problem. Actually, the first design thinking conception itself roots in an extracurricular course for mechanical engineers. Therefore, we advocate taking a step back and returning to the original idea of design thinking nowadays: the training of mechanical engineers to adopt the user perspective, which has substantially grown in popularity over the last decade (HPI Academy, 2020).

Why design engineers should become design thinkers

Beyond the processes of problem definition and solution finding, as highlighted by Peter Gorb and Angela Dumas (1987) in the late 1980s, "Design is also concerned with use, with marketing and production considerations and a wide range of technical and engineering resources and requirements".[5] Today, driven by increased levels of product complexity and interdisciplinary collaboration, design encompasses even more activities that traditionally were not considered part of designing. These involve, for example, activities to determine requirements, to set general directions, to evaluate design, and to observe in the field on earlier product usage situations or how a product is used after it has been released into the user's world. The Design Ladder (2001), for example, describes the shift of the role of design from product-focus to design integration on a strategic level: design as thinking provides a unique tool for approaching and solving problems.

Characteristics of user-centred design engineering practice

In order to realise a real user-centred design process, not only a user-centred design attitude is important, but also a constant generation of user information throughout the process. If one understands users as experts of their specific life-situation, it becomes evident why a participatory approach seems to be the most promising: both expertise, the one of designers, being experts in the development of tools, and the one of users, being experts of their experiences, are in demand when it comes to idea generation and concept development.

Although engineering designers generally confirm the importance of asking and observing what users require, there is discrepancy between the awareness of its theoretical importance and its practical

---

[5] The British academic Peter Gorb (1926–2013) is regarded as a pioneer of design management through his work at London Business School where he ran the Institute of Small Business and was among the first to teach design to managers. Angela Dumas, British Associate Professor at LBS, is the Founder & Non-Executive Director at Visual Metaphors at Work, UK, since 2012.

implementation (Zeisel, 2006). One reason for this lack of implementation lies in the difficulties engineering designers experience in collecting, selecting, understanding user needs and/or translating these into design language (Boztepe, 2007; Salmen, 2001). Another reason is that designers instinctively focus on designing for someone possessing physical skill capabilities without contact with target users with different capabilities (Dong et al., 2011). As it means remaining unaware of the users' actual needs, we see this attitude as very crucial Moreover, even if engineering designers do interact with users, they often do not know how to integrate the information gathered into the design process: they need data, which contains all the necessary elements adequately without information overload (Keates and Clarkson, 2003). The narrative of rather soft formulated qualitative user wishes totally differ from the representation of technical specifications engineering designers use to communicate and work with. Even talking to representatives of various target groups, aimed at revealing users' wishes as top priority, totally differs from the talking to, colleagues, for example.

Communication with and understanding of individuals or groups in different social contexts is not yet a relevant part of the standard curriculum in engineering education. Nevertheless, it is a relevant part in engineering design. Beyond technical competency, being an engineering design professional requires the capacity to address unstructured problems, to collaborate effectively in teams, and to have a high level of adaptability (Hora, 2017; Kind et al., 2019). We feel a strong need to emphasize this because, often, we receive industrial feedback that claim lacking capabilities particularly in the softer rather than technical domains such as practical working skills, creativity, applying research methods and communication (IET, 2015).

Consequently, when it comes to user-centred design and development, not only technology-driven perspectives and process models but human beings too must play a role in engineering education. Looking beyond one's own nose can help to extend their horizon in such way that truly innovative collaborations and concepts can emerge. This does not necessarily mean that one has to keep reinventing the wheel, and, in this sense, it definitively helps knowing whom to ask. The humanities, for example, not only offer theoretical concepts towards essential aspects that are necessary for a holistic understanding of users within their living context (like, e.g., social background, biography, lifestyle, motivation for acting, routines) but also research instruments to empirically access these (Wallisch and Paetzold, 2017).

Likewise the products themselves, with design engineering of such growing complexity, balancing the needs of multiple stakeholders requires the utilisation of a multitude of knowledge areas. Getting better is recommended to be prepared.

Characteristics of the design thinking way of thinking

Design as a value creation method aims at short- and long term outcomes for stakeholders. According to Tom and Dave Kelley (2013), feasibility, viability, and desirability, the three core characteristics for evaluating the success of a product or service, are defined as being "about deeply understanding human needs, […] understanding why people do what they do, with the goal of understanding what they might do in the future".[6] The message behind this is to staying focused on the people one is designing for, and listening to them directly, helps creating optimal solutions that meet their needs.

According to the global design and innovation company IDEO (2020), bringing together competencies in what is desirable from a human point of view with what is technologically feasible and economically viable turned out a good starting point to mix up teams assigned to designing solutions for people first.[7] When seeking solutions for people, the product's effect that engineering designers generally anticipate is to assist users in coping with their everyday life, to bring them joy or to enable a specific act first (for which air traveling probably serves as the most prominent example). Considering this, due to the heterogeneity of people in general and within particular user groups, it becomes clear why heterogeneously composed design teams are most likely to succeed in understanding who their users are, getting in touch with them and exploring their needs and validating the derived product requirements.

Design thinking is essentially perceived through the famous double-diamond process model of repeated explorative (divergent) and selective or defining (convergent) thinking steps (see Figure 1). Divergent phases cover the exploration of user needs and empathy building (discover) as well as solution generation (define), initial prototyping (develop) and testing (define), respectively. Convergent phases focus on the sense-making, selecting and defining, target outcomes for subsequent steps, the overall design aims or, eventually, the final design.

Usually, there are frequent jumps and iterations between phases, however, as insights generated at any stage in the process may affect

---

[6] The American executive and designer David Kelly (born 1951) is the founder of Silicon Valley-based IDEO and Stanford d.school creator. His brother Tom Kelley (born 1955) is partner at IDEO, developer of hundreds of innovative products from e.g. the first commercial mouse to virtual reality headsets and the Palm hand-held and the author of the bestselling 'The Art of Innovation' (2001).

[7] IDEO, founded by David Kelly who also founded Stanford University's Hasso Plattner Institute of Design (HPI), known as the d.school, is worldwide famous for being design thinking pioneer with roots dating back to 1978 in Palo Alto, California; since 1991 calling themselves IDEO. The claim to "always designing solutions for people first" is used at the IDEO (2020) website to highlight a user-centric mindset as the key tenet of design thinking, not matter what degree of complexity upcoming challenges require or how new methods developed to addressing these will look like.

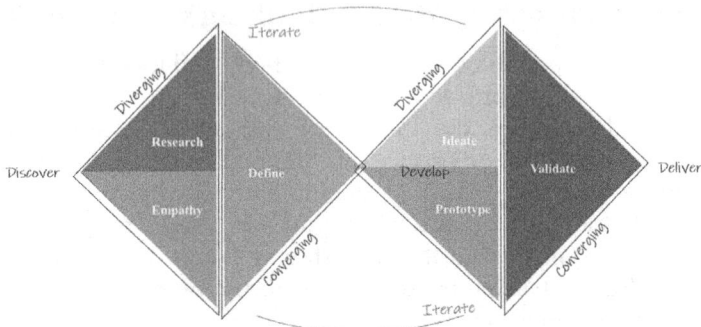

Figure 1. Design thinking double-diamond process model adopted from Brown (2009).

prior and subsequent stages. Plenty of methods and tools are available to formally support all involved steps and activities. A key characteristic is its clear focus on user-centricity, repeatedly—or even continuously—involving target user groups and other relevant stakeholders from multiple disciplines along the entire process for input, feedback and co-creation.

Design thinking iterates between user perspective (problem space) and product specification (solution space). An essential part of this iterative learning process is to build numerous prototypes, addressing various objectives in terms of exploration, communication, usability and design (Hallgrimsson, 2012). Despite playing such an essential role, it is not always necessary to build prototypes that reflect every function of the final product. Engineering designers in a highly uncertain and volatile environment are advised rather to build prototypes focusing on just a single or only a few design aspects. The overarching goal is gaining reliable insight in what serves the users most, validated by themselves. By building many small and simple prototypes, the development team can reduce risks early. Depending on how difficult it is to communicate information required for evaluation, prototypes do need to reflect neither a large variety of functions nor a big similarity to the final product.

In line with Jeanne Liedtka and colleagues (2017), according to which design thinking is understood best as a skillset such as the ability to handle uncertainty, tolerate ambiguity and maintain the big picture through systems thinking and systems design, we perceive design thinking as powerful instrument to developing these kind of capabilities that seem currently missing in engineering education.[8] To us, constantly

---

[8] Jeanne M. Liedtka (born 1955) is an American strategist and professor of business administration at the Darden School of the University of Virginia, particularly known for her work on strategic thinking, design thinking and organic growth. Randy Salzman is a journalist and former communications professor. Daisy Azer is an entrepreneur, principal at Waterbrand Consulting Inc., and adjunct lecturer of design thinking at the Darden Graduate School of Business.

observing how challenging interdisciplinary collaboration can be, it appears at least equally important to educate engineering students in successfully reflecting and communicating on their technical solutions as in technically designing them. From our point of view, this means nothing but encouraging them to evolve from being design engineers to becoming engineering design thinkers.

Course of action

The perception of design thinking as a systematic, human-centred strategy for solving complex problems within all aspects of life goes far beyond traditional concerns such as shape and layout. Design thinkers seek stepping into the users' shoes, which means not only interviewing them but also carefully observing their everyday behaviours. Concretizing and communicating idea and solution concepts as early as possible enables potential users continuously to provide valuable feedback on a product long before its finalization. Following this approach significantly minimizes the risk of developmental failures.

Doing design thinking

The design thinking process follows iterative loops guiding the design team through six phases (see Figure 2): in the first phase, understand, the team elaborates the problem space corresponding the design challenge given. In the second phase, observation, all team members gain an outward view and form empathy for their targeted users and other stakeholders involved. The third phase is about reframing the challenge by collating and summarizing the knowledge gained and, thus, to define the point of view. In the fourth phase of ideation, the team subsequently generates a variety of solution possibilities, and then selects a focus. The fifth phase, prototyping, is about developing concrete solutions, which are ready to be tested on representatives of the target group (sixth phase). A variety of professional backgrounds and functions, plus curiosity and openness for different perspectives, are the foundation of the creative working culture of design thinking. In design thinking workshops, a coach accompanies each team. He leads the team members through the entire process so they can focus on the contents of their constructive collaborative work

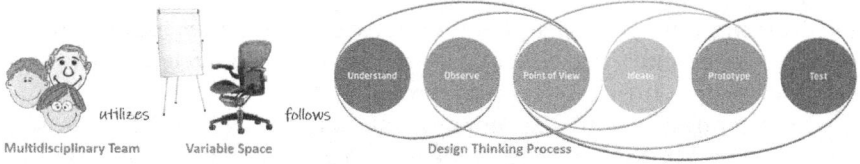

Figure 2. Core elements of the design thinking process after HPI (2020).

and reach their targeted goals. Teams constantly strive toward achieving tangible and concrete results by regularly exchanging with other teams to maximize the learning effect. Splitting up into small groups ensures a proper consideration for every perspective. A strong cohesion develops within the teams with a lasting effect, due to the high acceptance for the resulting concepts.

Design thinking teams work best in variable spaces, for example, standing up in spaces particularly designed for up to six people and allowing easy interaction with other teams working in parallel. A team needs optimal spatial conditions so that it can develop its creative process. These include flexible, movable furniture, adequate space for whiteboards and presentation surfaces as well as materials for prototyping design ideas, such as bricks, fabrics and images. This follows the idea of flexibility triggering creativity and out-of-the-box thinking, whereby working collaboratively is becoming a dynamic experience for everyone involved.

Design task and background

Designing technical systems is not always intuitive and even if so; sometimes designers have something different in mind as their users do. On the other hand, sometimes users feel disappointed by a product even if, from the design point of view, engineers did a very great job. Unsurprisingly, a number of product failures testify that engineering designers' assumptions on what particular user groups may like or not like are often quite wrong. Even if the same vocabulary is shared, people do not necessarily share the same meaning behind the words. As a result, information gets lost. In the worst case, the whole essence of a user's need gets lost in translation by engineers and designers; and she or he cannot grasp the potential of a design solution anymore. Both situations probably will lead to the product's market failure.

Although, on a cognitive level, an average person can estimate how aging and disability could affect another person, however, until he or she has dived deeper into that person's reality, their understanding is limited. Acquiring user-empathy means moving from a felt sense, which means to intellectually appreciating the experience of another, to a felt experience, which refers to the personal understanding gained when experiencing the experience of another person's experience. We see a strong need to foster engineering design novices in understanding these experience gaps and learning how to bridge them. Otherwise, guided by the traditional teaching in engineering education, they will probably tend to design solutions suiting their own scopes of reference.

User-centred design concepts in the tradition of usability testing consider only limited activities of the design process, namely those taking place before and after the actual product design. Design thinking, however, is about joint creation of value through adding up both expertise, that of designers and that of users. Given the fact that this perspective is lacking attention even in our very own academic environment, we discussed different options for pilot projects.[9] Finally, we decided on adding a design thinking part to the lecture on 'Methods for Product Development', which offers a case-based learning experience, specifically designed to enable undergraduate mechanical and aerospace engineering students to experience the challenges of older people, and its effects on everyday life.

Design thinkers pose many questions to potential users and take a close look at their processes and behaviour. Given the fact that the life experiences of young engineering designers and older people are usually different, this constellation gives an illustrative working example. Moreover, it allowed us to reuse materials from previous projects. For example, the needs of older people have already been surveyed and meaningfully clustered. The resulting incident cards served the present project as a source of core characteristics and typical scenarios to describe the target group. The students were about to choose three cards and derive the requirements from these. The development task was an adaptation development based on these three cards. In addition, we provided the students with a comprehensive collection of typical everyday scenarios older people perceive being challenging, and first solution concepts. From this, the students could choose one that they wanted to revise based on the previously determined requirement criteria. While working on the question, and which action situation they want to support, the students were already in the middle of the design thinking process.

Course design

Concerning the conceptualization of the learning experience, we did focus on the three core elements of design thinking (see Figure 2). The mix of mechanical and aerospace engineering students and their mentors or coaches represents multidisciplinary teams. The first coach was a sociologist aiming at sensitizing the students for the life situation of elderly people and the contradicting impact factors on their needs. She represented the user perspective throughout the whole design process by providing feedback on every iteration's outcomes and provided methodological input on the subject of how to gather and process the

---

[9] Reviewing the module descriptions of all engineering courses of study at our university showed: courses in humanities are rarely integrated electives, engineering teaching is predominantly organised in classical lectures. Searching for innovative courses in the user-centred design context, we certainly found some often turning out as being limited to guest lectures, role-play or cooperation projects aimed at the acquisition of students as future employees. Real project-based learning experiences in the sense of that the results are to be placed on the market does almost not appear at all.

user information actually needed. The second coach was an Industrial Designer aiming at providing the students with basic prototyping skills and theoretical knowledge as well as practical insights on the subject. A seminar room with movable interiors and writable walls served as 'prototyping lab' with different working materials the students could have chosen. This room was accessible to the students during the whole design thinking process and represented the design thinking success criterion of variable working space.

During the practical part, the students were intended to use and thereby train to apply different methods, concepts and tools they were introduced to during the lecture. They spent additional time on research through suitable options of data collection and adapting all this to their actual design process to ensure project success, motivated by the invitation of user representatives to the presentations of the project results. Figure 3 depicts a summary of the course design:

We conducted the outcome assessment of the design thinking process as a moderated focus group discussion session with representatives of the targeted users, organized in four similar rounds: starting with a team's presentation of their prototype, proceeding with its evaluation by the user representatives, ending in an open discussion on further improvements. Moderated feedback sessions may positively affect the participants' engagement with follow-up questions and, thereby, the clarification of confusing parts of the prototype, so we prepared an evaluation guideline to stimulate the discussion (see Figure 4).

For users, or their representatives, it could be really challenging to figuring out the questions they need to ask when testing a prototype. Moreover, the participating users sometimes tend to take an introduced design as given and feel neither confident nor responsible for major improvements.

Prepared with the evaluation aspects we previously defined, it was now on the participating elderly to give feedback on how well the students did actually achieve desired communication goals from their

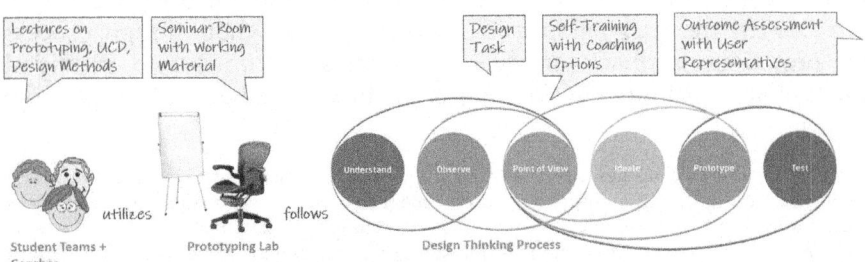

Figure 3. Core elements of the design thinking process after HPI (2020) adopted to the students project.

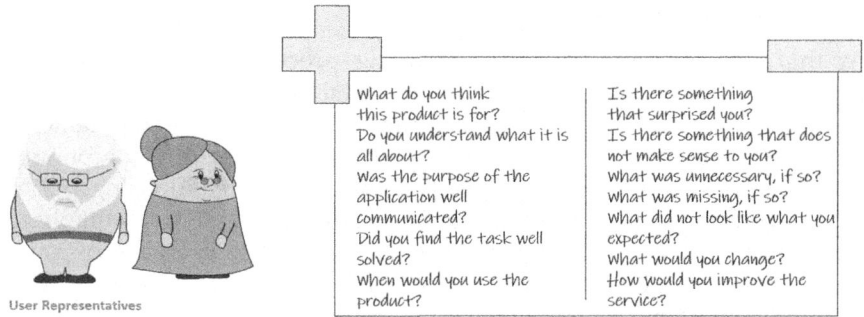

Figure 4. Guideline towards the assessment of the user representatives.

perspective (assessment). Providing not only feedback on the product concept itself but rather feedback on its presentation, we perceive this kind of assessment experience as an outstanding opportunity for project-based learnings including general design skills and useful implications, on how to improve user-centred requirements analyses and communication with non-designers in particular (closing the loop). To validate this perception, we provided a short questionnaire, which was passed on to the students at the end of the focus group session. Related to this, after a couple of weeks, we initiated a reflection session on the design thinking learning experience, whose discussion results will be discussed within the following.

Thinking design and proving design thinking innovations

Working for new or previously underexplored groups of people can be challenging. Nevertheless, it provides the chance to reflect on one's very own assumptions and long held beliefs in terms of good practices, thinking routines, stereotypes or the way of choosing and using design methods. Involving user representatives as design thinking partners in the design process guided our students in grasping the social role of design and its impact on people's quality of life more precisely. It also contributed to the students' perception of the value of designing. Designing an object with respect for someone's needs is motivated from a different angle in comparison to designing for being 'hipsterish'. The highest impact of having the user perspective—either represented by the coaches or the user representatives in the focus group session—involved thoroughly, was seen in the constant revision and discussion of the concept ideas, finally reaching a balance between the elderly' and the students' visions until clear problem statements were achieved.

Outcome assessment

At the beginning of the design thinking process, most students did appear somehow overstrained with the design task. By reviewing the incident cards, the students faced difficulties in deriving requirements from these. The translation of psychological, social, health and economic needs concerning various aspects of everyday life, such as housing, public transport, finance, safety, house chores, leisure, diseases, was quite hard for them, as this was their first convergence with both the target group and the usage context. Educated to think of rather technical requirements than qualitative user needs and not being familiar with core characteristics, they had no clear idea or routine on how to solve the problem properly. Educated in a rather traditional understanding of prototyping in its purpose of usability testing, implying testing of high-end prototypes in the late phases of the development process, it turned out to be more difficult than expected to inspire the students to use the material provided in order to create physical prototypes instead of digital ones. Continuously supported by the coaches during the design thinking process in all its iterations, the students eventually succeeded in acquiring a more user-centred mindset, allowing not only to address user problems technically but also to assess solution concepts from the targeted users' perspectives.

At the beginning of the focus group discussion session, the motivation was quite high and both sides, students and user representatives, seemed very excited about what would happen. At the end of the session, both sides seemed deeply impressed by the contributions made by the respective other. The students did use various prototypes to present their solution concepts: Team 1 presented a shopping cart with step by step oral explanations of 3D models done with CATIA software and technical sketches.[10] Team 2 did use SAP Scenes to introduce their product service system and presented an exemplary user journey through the neighbourhood-based leisure services they offered.[11] Team 3 presented their idea of height-adjustable kitchen cupboards in the form of an advertising video with oral explanations. Eventually, Team 4 presented the "Steakholder", a low fidelity prototype of their solution to assist

---

[10] CATIA (an acronym of computer-aided three-dimensional interactive application) is a multi-platform software suite for computer-aided design (CAD), computer-aided manufacturing (CAM), computer-aided engineering (CAE), PLM and 3D, developed by the French company Dassault Systèmes.

[11] SAP Scenes is a tool to create storyboards about products and services fast, collaboratively and iteratively. It empowers to shape one's ideas and scenarios in the form of illustrative stories without the need of refined drawing skills. Scenes includes a set of pre-defined illustrations to create a visual story (physical and digital). Link to download free kit in References (SAP, 2020).

people who are restricted in the gripping behaviour of their hands, in the course of a role-play.[12]

To understand and test prototypes in the intended usage context, their presentation should reflect the situation and people involved. The prototype's clarity and an understanding of the product are important elements for users to understand the product prototype in such a way, that they can comment on it to provide designers with valuable feedback. For older aged people clear and simple prototypes they already are familiar with from other contexts appeared to be the most suitable because they already are experienced in interpreting these kinds of objects. The user representatives confirmed that all solutions are relevant to elderly peoples' life situations and address the respective issue appropriately. They enjoyed the different presentation styles, all of which they perceived as explaining the respective solution concept very well, but particularly highlighted the opportunity of trying as given by the low fidelity prototype. The students, on the other hand, were surprised on how differentiated the feedback given by the user representatives was and believe that the final design did significantly improve by the user's input, and the final solution has great usability. The majority of students perceived the interaction with the user as consistently positive, applying both experiences in communication as well as in collaboration.

Summary of key learnings

Not surprisingly, even if the students were highly motivated at improving the elderlies' everyday life, they firstly focused on technical feasibility. Consequently, in the beginning of the design process, they were looking for high-end solutions, which were too complex to meet the elderlies wish for rather frugal and low fidelity solutions, which were perceived as being more easily includable in their daily routines. This difference, of thinking in frugal solutions or high-end solutions, reflects the main competence areas of both users as experts of their life situations and designers as experts of solution finding, and thereby gives reason for direct interaction with user representatives throughout the development process.

Continuous prototyping minimizes both the communication—as well as the experience gap between the potential user and engineering designer. As it promotes constant improvement with every iteration, it serves as a powerful strategy to develop user-centred designs, which we perceive as not well reflected in engineering education. The students are just used to either virtual models, which are due to their complexity not appropriate in

---

[12] A summary with short descriptions of all technical solutions and pictorial impressions of the prototypes presented as well as feedback protocols is to be found within the appendix of this chapter.

terms of effective user communication or even collaboration, or high-end prototypes, which need great effort in changing if an issue emerges. They were neither aware nor had the least idea of the potential of prototype-based communication, especially in the very early development phases, because they did not have the chance to practice it in their previous education.

Within their reflections on the aspects of life situations based on which the students developed their concepts, they typically referred to 'challenges' of seniors compared to younger people. Although attempts are made to do so from the users' point of view, this is more likely to follow a deficit-oriented perspective, which at some points leads to bad communication of good ideas: The user representatives were not amused being stigmatized as people who need special assistance and have no idea of the latest technical and social innovations. Closely related to this, during the design thinking process, at some point the coaches observed an interesting change of perspective happening: from concept development to concept detailing, within the teams, as the user focus decreases. As students were asked, why they eventually decided on one solution, the answer mostly was like "then I had an idea, which I liked much". Continuous interaction with user representatives provides fruitful feedback to validate these.

The observations students made with their own grandparents and neighbours maintained comparatively little feedback on their impressions from the user group itself, so observations could also be wrong and/or biased. Moreover, neither their own family members nor the invited users represent all kinds of older users belonging to this heterogeneous user group. A complex aspect of design is the need to consider multiple variables, which forces engineers to get to know diverse issues being relevant to the product or service to be developed. A design solution may be virtuous for some, but negative for others. Responsible evaluation of these aspects by referencing to diverse areas of knowledge requires research activities, which include collecting and contextualizing the essential information needed to end up with an appropriate product or service finally. Students should have the possibility of taking courses in questioning and exploring as well as in assessing user information from different resources.

Finally, the user representatives did show some stereotypes towards engineers, too: One participant cited a report thematising "the speechlessness of engineers" and another confessed to wondering what relation exists between a service solution and a technical product. It is important to be aware of these kinds of experience gaps and diverging expectations leading to double-biased communication mostly. Within our presentation session, communicational validation turned out as being the most important aspect when it comes to an actually user-centred development of user-centred design solutions. A prototype's presentation that included oral explanations were highlighted as well understandable in general and particularly if given in a language quite close to the elderlies' own language expertise.

Best practices for carrying out actions and activities

Engineers sometimes do not see any effect of changing their traditionalized thinking routines because every technical system they develop belongs to a customer determining the requirements. They are not even aware of the leading role technical feasibility plays in their working process and how it affects all design decisions. To catch the engineers' attention on design thinking, it turned out fruitful reminding them of a specific definition of innovations, whereas innovation is an invention being bought (Schumpeter, 2008). The central question now is "who buys?", and the answer is quite simple "it's people". Consequently, the design thinking strategy is to start with the people, believing this will solve all other issues during the development process anyway.

As this is contradicting the traditional approach they are usually following, engineers may face similar difficulties in transferring the design thinking steps into their design practice as was observed with the engineering students in this pilot course. The experiences made within this pilot, however, can be condensed into measures supporting the design engineers' decision-making within each single design thinking step in terms of its how to be carried out. The following guide through design thinking activities roughly depicts our key takeaways (Figure 5).

In terms of understanding, it is worth trying to forget all well-known patterns in order to understand the things seen from within themselves. One premise is to consider as many different perspectives as possible. When it comes to observing, it is essential to understand the reality of the problem to be solved, e.g., through interviewing representatives of the targeted users and television reports. Then maybe the most powerful

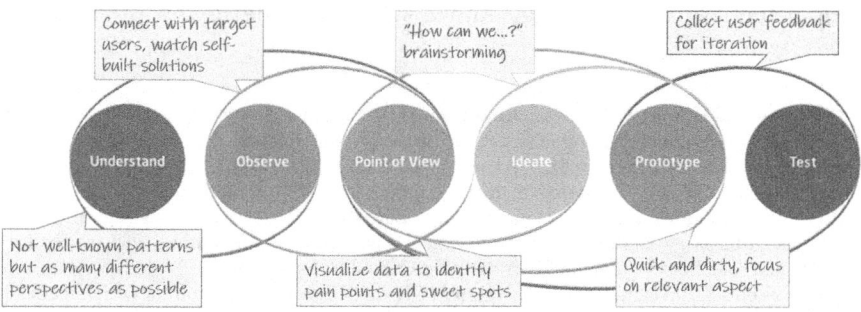

Figure 5. Design thinking pocket guide.

approach is to keep an eye out for self-built solutions, because these not only reveal a real need but at the same time help to understand the user's thinking and gaining empathy for them. User-made prototypes come from the very own sphere of experience and knowledge and therefore contain the essence of the problem's solution. As far as defining is concerned, it is helpful to use visual work such as, for example, differently coloured post-its to find the patterns lying in the collected data with greater ease. Describing the steps of the user journey using interview data and matching it with the pain points supports in identifying the sweet spots. Sweet spots show up at steps at which as many pain points as possible are concentrated and, at the same time, these steps are the ones open to change with relative ease.

To fostering ideation, it is effective to implement question-based brainstorming, starting with "how might we...?", for example. Efficient teams generate a very high output in a very short time. In this stage, the focus is on generating ideas only, whereas the idea's evaluation is not a part of it. The most powerful method to test ideas is implementing prototypes as simply and as quickly as possible. Sometimes, prototypes are only aiming at communicating an idea; in this case, they do not necessarily have to prove they are already functional. On the other hand, function does not necessarily have to be beautiful. If chosen wisely, the combination of function-, design- and show prototypes achieves the most valid results. The Media Richness Theory proposed by Richard Daft and Robert Lengel (1983) offers assistance in choosing the right prototype corresponding to the communication task given (Schmidt et al., 2017).[13] Finally, it is obvious to bring the prototype(s) developed quickly to its targeted users and, based on the testing, into iteration.

Reflections on the findings and implications for engineering education

Within our own academic environment, keep-it-simple-prototyping is barely considered relevant to mechanical engineering students. As people tend to do what they know, when it came to the prototyping

---

[13] Richard L. Daft (born 1941) is an American organizational theorist who developed and managed the Center for Change Leadership at the Owen Graduate School of Management, Vanderbilt University, where he also served as Associate Dean for Academic Programs. He is a Fellow of the Academy of Management. Robert H. Lengel (born 1946) spans a career that includes aerospace engineering, industrial sales and marketing, financial services and money management, environmental analysis and consulting, and academics. In this last role, he served as the founding director of an Executive EMBA program and an executive education centre at the University of Texas at San Antonio, both of which focused on the creation and application of innovative transdisciplinary approaches to leadership development and change dynamics.

part, the mechanical engineering students started creating CAD-models mainly, just out of habit. The students did not know how to use most of the working materials provided and they did not expect the effect that physical prototypes could have on design communication. Revealed as powerful communication strategy, the (missing) role of prototyping in engineering education deserves a second glance.

During the design thinking process, this group went to the coaches, asking if their prototype "will really be enough" or if they additionally have to build a CAD model for the final presentations. During the feedback session, the invited older ones confirmed that CAD models were too complex to communicate simple aspects of a concept prototype. This tendency to overcomplicating technical solutions is an outcome of academic education, which should preferably not be the case. In contrast, engineering education has to prepare students for the constantly changing demands of the market, so it seems advisable to, understand prototyping used as a design strategy, and given a higher priority in education and utilize makerspaces as a resource for user-related requirement elicitation rather than just specific product iteration (Jensen and Steinert, 2020).

Considering this, the design thinking pilot turned out as being perceived extremely valuable not only in terms of the design solutions developed but also with regard to the whole design process and in particular the aspects of collaboration and design communication. Regarding the question, of how they perceive their own learnings, the majority of students reported that participating did help them to grasp the meaning of requirements analysis more holistically. Some students also mentioned that their awareness of the interdependency between communication media richness and the communication task went to a higher level and others highlighted "skills in prototyping" as being another valuable competence acquired within this pilot course in design thinking.

Within the reflection, all students agreed on that their initial ideas were highly complex because they tried to solve multiple needs of the older people. The design thinking experience activated and sharpened the students' design skills, and the interaction with the user representatives allowed them to move from complex to rather simple solution concepts. This interaction enables empathy for targeted users and therefore drives a shift in thinking design: from a predominantly technology—to a more user-driven perspective. After the focus group discussion session, the students reported that the direct interaction with the seniors did change their whole perspective on the purpose of design solutions, in particular the requirements analysis, and confirmed this experience did sharpen their understanding of what it means to be a design professional (Baha et al., 2020; Dunne and Martin, 2006).

As empathizing with users is a critical success factor in new product development, from requirements elicitation to market launch, facilitating user-empathy skills should be a central learning goal in engineering education. Bringing engineering students closer to users should become an academic teaching task. Dedicated learning experiences and curricula for those, who are designing and need to know how best to function in the generative front end of the product development process, are required. In addition to the practical design skills, our participating students soon developed a high degree of empathy for the targeted user group that was initially unknown to them.

Another aspect, which we observed during this course and got the impression of making it a key aspect of education in the future, targets the criteria determining the transformation process from user needs into technical specifications. When asked about what they finally decided on a concept, some students answered, "Because I liked it". This answer shows us potentials for educational goals, which obviously need to be supported more strongly: Either the students simply find it difficult to communicate what they have decided on because they cannot grasp their intuitively made decisions properly, in which case engineering education must methodically support them in their decision-making process. On the other hand, if the answer indicates that the students simply did not think about it properly, i.e., lost the user perspective as the development process progressed, we should support them more in reflecting on what they are doing. Within our teaching experience, design thinking showed potential for improving both.

Learning objectives in engineering education being successfully targeted by design thinking

Based on the learnings presented, students would benefit from design thinking becoming a standard part of the curricula in academic engineering education in terms of

1. Improving general learning outcomes

    As it provided the opportunity of practical training of various design skills and methods, the students rated the whole design thinking experience as useful for their future professional and even educational process. Engineering education in general should focus more on teaching formats focusing on applying knowledge instead of just presenting it.

2. Providing skills as demanded by industry

    The students declared having improved their personal social- and communicative competencies as well as their activity- and implementation-oriented methodological and professional competencies during the design thinking process, as they were able to acquire knowledge according to the principle of learning by doing and thus consolidating it at the same time.

3. Preventing from design fixation

    Having user representatives involved in the design process allows instant validation of all identified requirements and, thereby, concept development and detailing according to their input on what they need instead of what engineers imposed on them. By constant interaction with both the coaches and user representatives, the students felt forced to reflect their own biases continually in terms of existing issues and practices.

4. Improving design solutions

    Functioning as an enabler for students' thoughts, prototypes allow better communication amongst the project's stakeholders, such as students, teachers and potential users and support students in tackling challenging problems. Physical prototypes and models in particular proved to be more beneficial in communicating specific aspects of design concepts than other forms of representations such as technical sketches, and helped to find innovative ways to solve complex design tasks gradually.

5. Unlashing new design mindsets

    Design thinkers are approaching problems through the eyes of the targeted users by putting themselves in the user's role. Understanding the targeted users helped the students not only with regard to requirements engineering, but above all, with a design attitude that allows integrating the elderlies' perspective into the ongoing design process continuously.

6. Minimizing the risk of developmental failure

    Continuous prototyping from the very beginning makes concept ideas and solutions accessible and communicable as early as possible so that potential users can test these and provide feedback long before they are completed or launched on the market. Even complex systems can be broken down into individual communication tasks and solved with the help of low-fidelity prototypes, minimizing the risk of failure and reducing complexity.

Current literature provides a plethora of definitions for design thinking with variations in tools, methods and cultures. We follow its meaning as a continuous scale of practices, cognitive approaches and mindsets aiming for stronger user-centrism in organizations and projects. For the sake of completeness, it should be noted that design thinking is not suitable for solving every question, such as, e.g., the increase in performance of a machine is not necessarily achievable through interviews with targeted users. From our point of view, when it comes to design thinking in engineering education, this is more an additional advantage than being a contradiction: teaching a design thinking mindset means teaching people to understand the structure of a problem, recognizing a lack of knowledge on their side, offering creative solutions on how to overcome it, and, thereby, fostering appropriate decisions.

Closing remarks

Academic teaching of engineering students comes along with the responsibility to educate the next generation of product developers, which means it has to react to different trends like increasing interdisciplinary collaboration and participatory design to enable students for lifelong learning in their field of expertise. The evolution in design paradigms, from user-oriented to participatory, accompanied changes in the way we perceive the roles of engineers, designers and users, placing new demands on engineering education for product developers.

Considerations on engineering education should include two things: First, all engineering education is the result of educators and their framing of good engineering, forming the basis of curricula, courses, pedagogy, and educational models. Second, student's perception of engineering design rests on these frames, influencing them in becoming a design engineering professional. Prior to starting their education program, most students miss a clear perception of engineering in general and the particular design philosophy they are subscribing to, meaning preferences towards specific perceptions of engineering design are set by their choice of school. Identity development as a professional starts as soon as students enrol into an undergraduate engineering program, gradually developing an authorized understanding of engineering, and learning to meet prescribed standards.

Overcoming the conservative engineering approach of solely thinking of technical solutions, design thinking is to be understood as an innovation or entrepreneurship craft offering a toolkit, which needs to be taught and trained well in order to being pursued successfully. The continuously increasing number of courses and programs each of varying style and format, aimed at qualifying professionals and non-professionals with any background in design thinking, proves an existing educational need. High quality training formats aiming at actually stimulating and establishing sustainable change in one's design thinking behaviour are expensive and time-consuming. To prevent academic education from risking missing out an enormously relevant demand for graduates with an appropriate mindset and skills corresponding to their upcoming tasks, we advocate design thinking becoming a standard part in engineering education curricula.

References

Baha, E., Koch, M., Sturkenboom, N., Price, R. and Snelders, D. 2020. Why am I studying design? pp. 1898–1915. In: Boess, S., Cheung, M. and Cain, R. [eds.]. Synergy—DRS International Conference 2020. 11–14 August, held online.

Bayazit, N. 2004. Investigating design: A review of forty years of design research. Design Issues, 20(1): 16–29.

Bouwman, S., Voorendt, J., Eisenbart, B. and McKilligan, S. 2019. Design thinking: An approach with various perceptions. Proceedings of the 22nd International Conference on Engineering Design (ICED19). 5–8 August. Delft. The Netherlands.

Boztepe, S. 2007. Towards a framework of product development for global markets: A user-value-based approach. Design Studies, 28(5): 513–533.

Brown, T. 2009. Change By Design. HarperCollins Publishers. United States.

Carlgren, L., Elmqvist, M. and Rauth, I. 2016. Exploring the use of design thinking in large organizations: Towards a research agenda. Swedish Design Research Journal, 11(1): 55–63.

Cross, N. 1982. Designerly ways of knowing. Design Studies, 3(4): 221–227.

Daft, R.L. and Lengel, R.H. 1983. Information Richness—A new Approach to Managerial Behavior and Organization Design. Defense Technical Information Center. Fort Belvoir. VA.

Danish Design Centre 2001. The Design Ladder. Danish Design Centre, online. URL: https://danskdesigncenter.dk/en/ design-ladder-four-steps-design-use. Accessed 1 September 2018.

Dong, H., Mcgingley, C., Nickpour, F., Chen, H. and Pei, E. 2011. Evaluating inclusive design tools: An insight include 2011. pp. 1–10. Helen Hamlyn Centre for Design, London.

Dunne, D. and Martin, R. 2006. Design thinking and how it will change management education. Academy of Management Learning and Education, 5(1): 512–523.

Goldschmidt, G. and Badke-Schaub, P. 2009. The design-psychology indispensible research partnership. Proceedings of the 8th Design Thinking Research Symposium. Sydney.

Goodman-Deane, J., Langdon, P.M., Clarkson, P.J. and Clarke, S. 2008. User involvement and user data: A framework to help designers to select appropriate methods. Designing Inclusive Futures. pp. 23–34. Springer. London.

Gorb, P. and Dumas, A. 1987. Silent design. Design Studies, 8(3): 50–156.

Hallgrimsson, B. 2012. Prototyping and Modelmaking for Product Design. Laurence King Publishing.

Hora M.T. 2017. Beyond the Skills Gap, National Association of Colleges and Employers, online URL: https://www.naceweb.org/career-readiness/trends-and-predictions/beyond-the-skills-gap/. Accessed 10 January 2020.

HPI Academy 2020. Education for Professionals: Design Thinking, online. URL: https://hpi-academy.de/design-thinking/was-ist-design-thinking.html, Accessed 14 March 2020.

Huppatz, D.J. 2015. Revisiting Herbert Simon's "science of design". Design Issues, 31(2): 29–40.

IDEO. 2020. About IDEO, online. URL: https://www.ideo.com/about, Accessed 18. November 2020.

Institute of Engineering and Technology (IET) 2015. Skills and Demand in Industry Survey 2015, Institution of Engineering and Technology, online. URL: https://www.theiet.org/impact-society/factfiles/education-factfiles/iet-skills-survey/iet-skills-survey-2015/. Accessed 14 January 2020.

Jensen, M.B. and Steinert, M. 2020. User research enabled by makerspaces: bringing functionality to classical experience prototypes. Artificial Intelligence for Engineering Design, Analysis and Manufacturing, pp. 1–12.

Keates, S. and Clarkson, P.J. 2003. Countering design exclusion: bridging the gap between usability and accessibility. Universal Access in the Information Society, 2(3): 215–225.

Kelley, T. and Kelley, D. 2013. Creative Confidence: Unleashing the Creative Potential Within Us All. Currency.

Kimbell, L. 2011. Rethinking design thinking: Part I, Design and Culture, 3(3): 285–306.

Kind, S., Dybov, A., Buchholz, C. and Stark, R. 2019. Application of industrial methods on engineering education. 21st International Conference on Engineering and Product Design Education. E&PDE 2019. September 12–13th. pp. 1–6. University of Strathclyde, Glasgow.

Kosmala, M., van der Marel, F. and Björklund, T. 2019. Interpretations of design thinking across a large organization'. Proceedings of the 22nd International Conference on Engineering Design (ICED19). 5–8 August. Delft. The Netherlands.

Liedtka, J., Salzman, R. and Azer, D. 2017. Democratizing innovation fashionable organizations: Teaching design thinking to non-designers. Design Management Review, 28(3): 49–55.

Norman, D.A. 1988. The psychology of everyday things. Doubleday. New York.

Norman, D.A. and Draper, S.W. [eds.]. 1986. User-Centered System Design: New Perspectives on Human-Computer Interaction. Lawrence Earlbaum Associates. Hillsdale. NJ.

Rittel, H.W.J. and Webber, M.M. 1973. Dilemmas in a general theory of planning. Policy Sciences, 4(2): 155–169.

Salmen, J.P.S. 2001. US accessibility codes and standards. pp. 1211–1218. In: Preiser, W.F. and Ostroff, E. [eds.]. Universal Design Handbook. McGraw-Hill. New York.

Sanders, E.B.-N. and Stappers, P.J. 2008. Co-creation and the new landscapes of design. Co-Design, 4(1): 5–18.

Sanders, E.B.-N. and Stappers, P.J. 2014. Probes, toolkits and prototypes: three approaches to making in co-designing. Co-Design, 10(1): 5–14.

SAP 2020. Scenes - Every great experience starts with a great story, online. URL: https://experience.sap.com/designservices/resource/scenes?ref=publicdesignvault, Accessed 18. November 2020.

Schmidt, T.S., Wallisch, A., Böhmer, A.I., Paetzold, K. and Lindemann, U. 2017. Media richness theory in agile development—Choosing appropriate kinds of prototypes to obtain reliable feedback. International Conference on Engineering. Technology and Innovation (ICE/ITMC). IEEE. pp. 521–530. Funchal.

Schumpeter, J.A. 2008. Capitalism, Socialism, and Democracy. 3rd edition. Harper Perennial Modern Thoughts. New York. NY.

Wallisch, A. and Paetzold, K. 2017. User integration in design: Applying social science research questions for successful participation management at every product development step. pp. 39–48. In: Vajna, S. (ed.). 11th International Workshop on Integrated Design Engineering—IDE Workshop'17, Otto-von-Guericke-University Magdeburg, Germany.

Wallisch, A. and Paetzold, K. 2018. A qualitative inventory of user integration methods and their usage in product development research and practice. Proceedings of the 15th International Design Conference (DESIGN 2018). 20–24 May 2018. pp. 115–126. Dubrovnik, Croatia.

Wallisch, A. and Paetzold, K. 2020. Methodological foundations of user involvement research: A contribution to user-centred design theory. Proceedings of the 16th International Design Conference (DESIGN 2020). 26–29 October 2020. pp. 71–80. Cavtat, Croatia.

Wallisch, A., Sankowski, O., Krause, D. and Paetzold, K. 2019. Overcoming fuzzy design practice: revealing potentials of user-centered design research and methodological concepts related to user involvement. Proceedings of the IEEE International Conference on Engineering. Technology and Innovation (ICE/ITMC). 17–19 June 2019. Valbonne Sophia-Antipolis, France.

Zeisel, J. 2006. Inquiry by design. Norton, New York.

Appendix

Descriptions of the technical solutions and prototypes presented, and feedback:

Team 1: „Shopping cart with step"	
Problem description	elderly people are too small to reach products, which are at the top of supermarket shelves; they do not want to use extra equipment
Technical solution	add a small step on all shopping carts
Prototype	CATIA models and sketches
Presentation	PowerPoint
Feedback concept	nice idea, but maybe perceived as being too dangerous if not an extra holding option
Feedback choice of prototype	a technical model is not the best option to present the idea because of perspective distortion when integrating the model into a real life picture of the supermarket; too over-complex for the elderly, the sketches are too hard to follow for the elderly, without any connection to their real life living situation, you have to work on the geometry, something is missing to hold on to, the shopping cart step is not in order, problem of stigmatization
Feedback presentation	very good

Team 2: „Neighbourhood-based leisure services"	
Problem description	elderly people who are not any longer tied to their colleagues tend to get isolated from social activities
Technical solution	service system, community platform, messenger
Prototype	SAP Scenes and Persona
Presentation	PowerPoint and storytelling
Feedback concept	nice idea, but maybe seniors do not want to go on bus rides with "really old" people, you don't have to describe it as for elderly exclusively but still address them (maybe by the selection of times and places)
Feedback choice of prototype	very lively and close to the real life situation of the elderly, easy to follow, it was fun and created positive UX/emotions, pen and paper style was very nice, even if it was digitalised "where is the technical product, I have to evaluate?", "I did not know that technicians also develop services", product service systems are very appropriate to meet the target group's requirements
Feedback presentation	storyline was very good, with the exception of this "have you ever heard of tinder"-stigmatization in the introduction

244 Different Perspectives in Design Thinking

Team 3: „Height-adjustable kitchen cupboards"	
Problem description	elderly people are not so tall anymore, don't want to climb to get stuff down from the cupboard, they tend to leave a mess in the kitchen
Technical solution	result 1: continuous rails mounted on the wall – but (problem analysis): visually not appealing, conspicuous, too spacious, difficult to clean, backsplash result 2: hidden electrical system, spring-rope mechanism (analogous to height-adjustable desks): rails are hidden in the cupboard: not visible from the outside, space saving, cleaning of the guideways not necessary
Prototype	CAD-model, video
Presentation	PowerPoint with demonstration video and explanation
Feedback concept	very interesting for elderly people, it has to be some automatic stoppers included
Feedback choice of prototype	video clip was very lively and close to the real life situation of the elderly, older people tend to reduce their radius of action in the case of physical limitations and also lose important training effects through less exercise, as getting older it looks more chaotic in the household because everything should be within reach: so the flour stands on the worktop instead of in the cupboard, the solution presented here could be a remedy, it does even have a "coolness factor" because it looks very modern and fancy (reminds of smart home solutions), elderly were used to these kinds of advertising spots, so they could follow the message easily
Feedback presentation	particularly the combination of film and oral explanation was very good

Team 4: „Steakholder"	
Problem description	motoric abilities are getting less after strong diseases, people can't grab things like that anymore, they need support in training their motoric skills
Technical solution	a hand cuff with fixing tape
Prototype	low fidelity prototype
Presentation	role play
Feedback concept	very perfect for recovering and training reasons, "would like to have it in different colours"
Feedback choice of prototype	absolutely perfect, because it could have been tried and thereby, the idea got really clear, the best possible choice of prototype for the target group, colour and usability are sweet spots for user attractiveness
Feedback presentation	very good, very clear, just the analogy to children is perceived as somehow not appropriate because children are little persons one has to take care of

Index

A

Abstract problems 2
Accessibility 177, 181, 185, 187, 191, 200
Act of designing 117
Acting 43, 50, 52, 53, 57, 61, 111, 224
Action research 111, 112, 139–143, 156, 159
Actor Network Theory (ANT) 50
Ageing 202, 208, 209, 212
Akin, Ömer 12
Algorithm 133
Analytics 131, 153, 154, 156, 160
Artefact 67, 68, 72, 74, 84, 120, 132, 138, 140, 146, 150
Assessment 15, 16, 172, 193, 222, 230–232
Attention 7, 9, 11, 24, 30, 31, 39, 57, 68, 69, 71, 76, 82, 88, 89, 109, 119, 126, 136, 146, 186, 187, 189–191, 202, 229, 235
Awareness changing 6

B

Backlinks 12, 13, 15
Bardzell and Bardzell 7, 9
Beer, Stafford 143, 149, 151
Behavior 29, 35, 68, 70, 81, 83, 86, 93, 94, 101
Behavioural fit 153
Belting, Hans 79
Binary 42–44, 46, 90, 98, 100, 104
Biological sex 89, 90, 92, 93
Boundary objects 153
Brainstorming 2, 216, 236
Brown, Tim 16, 35, 41, 53, 88, 100
Butler, Judith 43, 98

C

C.critical T.thinking 6
Care manager 166, 176, 179, 185, 187, 190, 195, 198–200
Care service 165, 166, 169–171, 195
Caregiver 181, 182, 185, 189, 190
Case studies 28, 142, 155
Categorization 69, 84, 90, 176
Categorize(ies) 84, 203
CATIA 232
CATWOE 141, 157
Checkland, Peter 112, 141, 142
Co-creation 63, 171, 191, 192, 205, 209, 214, 217, 226
Co-design 50, 52, 54, 61, 69, 111, 211
Cognitive 3, 5, 7, 9–12, 29–34, 36, 48, 56, 68, 71, 75, 82, 85, 94, 134, 137–139, 153, 157, 158, 169, 193, 210, 221, 222, 228, 239
Cognitive level 11, 228
Cognitive map(ping) 137–139, 157
Cognitive scale, acts and process 10
Collins, Patricia, Hill 91
Communication 11, 31–33, 41, 45, 46, 51, 53, 56, 69, 71, 75–77, 79, 82–86, 90, 96–99, 104, 108, 115, 117–123, 138, 143, 149, 153, 156, 168, 169, 171, 175, 188, 192, 202, 203, 206, 211, 213, 224, 226, 230, 231, 233, 234, 236, 237, 239
Complex design 110, 117, 126, 239
Complex problem(s) 110, 115–117, 126, 131, 227
Concept development 223, 234, 239
Connoted, connotation 46, 56
Context 3, 16, 18, 22, 28, 31, 33, 34, 36, 42, 45, 48, 49, 54–56, 59, 67–69, 74, 76, 77, 79, 83, 85, 97, 101, 120–123, 138, 143, 144, 156, 166–168, 171, 175, 185, 188, 191, 193, 204–207, 209–217, 224, 229, 232, 233
Convergent phases 54, 225
Covid19 77, 109, 116
Critical path 7, 9, 10, 12, 13, 15, 16
Critical Thinking 6
Cross-generational groups 211, 215
Cross-group 212
Cross, Nigel 88, 111, 132, 221
Cultural 45–48, 51, 52, 55, 56, 59, 61, 62, 67, 71, 75–77, 83, 85, 89–93, 101, 105, 109, 156, 186, 194
Cybernetics 134, 143, 146, 147, 151, 157

D

D.design T.thinking 1–7, 10–12, 16–18, 21–29, 32–39, 41–43, 45–48, 50, 52–54, 56–63, 68, 69, 71, 73, 74, 77, 81, 83, 86, 88–90, 92, 95, 97–102, 104, 105, 111, 112, 116, 120, 127, 131–136, 138, 140, 142–147, 149–152, 154–160, 165–172, 192, 194, 195, 205, 207, 208, 220, 222, 223, 225–232, 234, 235, 237–240
Decision 15, 16, 49, 55, 95, 97, 131, 132, 135, 139, 143, 144, 152, 153, 156, 158, 159, 168, 171, 195, 235, 238
Decision making 49, 55, 135, 139, 152, 153, 156, 168, 173, 235, 238
Decision support 139
Define 5, 7, 9, 22, 23, 42, 44, 46, 59, 69, 93, 101, 139, 146, 151, 153, 156, 185, 225, 227
Denotative 76
Denoted 47, 48, 57
Design criticality 16
Design dialogues 55, 108, 110–114, 117–119, 121, 126, 127
Design education 3, 39, 47
Design fixation 239
Design for everybody 101
Design history 94, 116
Design Method movement 10, 11, 17
Design methodology 4, 5, 7
Design methods 5, 10, 11, 17, 34, 41, 61, 109–111, 114, 117, 143, 144, 204, 231
Design mindsets 239
Design of everyday objects 96
Design of products, spaces, services, system, 180, 121
Design practice 21, 37, 38, 42, 47, 48, 50, 53, 57, 58, 61, 62, 90, 94, 134, 222, 235
Design process 2, 5–7, 9, 10, 12, 13, 18, 24, 32, 34, 38, 39, 47, 50, 52, 55, 57, 59, 61, 83, 86, 92, 97, 108–110, 112, 113, 115, 117, 135, 144, 153, 202–206, 208–212, 214–216, 221, 223, 224, 228–231, 233, 237, 239
Design Research 1, 3, 5, 12, 18, 50, 59, 61, 68, 69, 86, 111, 126, 136, 165–168, 176, 221
Design sessions 209, 210
Design solution 22–24, 34, 83, 144, 228, 234, 235, 237, 239
Design studies 34, 169, 221
Design thinking approach 2, 26, 31, 36, 48, 54, 61, 147, 159, 170–172, 192, 205, 207
Design Thinking movement 10, 36, 38

Designerly 35, 61, 88, 111, 114, 131, 132, 134, 135, 137–139, 142, 144, 146, 151, 158–160
Development process 2, 108, 110, 111, 167, 209, 212, 222, 232, 233, 235, 238
Dewey, John 111
Dichotomies 50, 90
Differentiating 95
Digital transformation 165
Digitalization 102, 165, 202
Direct reasoning question (DRQ) 4
Disability 69, 166, 228
Disruptive innovation 2
Divergent phases 225
Divergent thinking 4, 22, 24, 26
Diversity 29, 59, 61, 62, 71, 92, 104
Dunne, Anthony 7

E

Easy-to-read language 69, 76
Education 1, 3, 6, 8, 21, 27, 29, 32, 34–37, 39, 47, 68, 77, 85, 117, 123, 167, 175, 192, 220, 222, 224, 226, 228, 233, 234, 236–238, 240
Embedded information 69, 72, 84
Empathize 54, 59, 69, 71, 77, 97, 104
Empirical studies 3
Empowerment 191, 208, 210
End-users 166, 167, 176, 178–181, 185, 187, 189–191, 195, 198, 199
Engaging 24, 33, 149, 188, 207, 209, 214, 215
Engineering designers 223–226, 228, 229
Everyday life 43, 165, 205, 212, 225, 229, 232, 233
Experience gap(s) 228, 233, 234
Exploration methods 111, 112

F

Female 42, 43, 45, 46, 89–92, 95, 98, 99, 104, 121
Femineity 89, 95, 99
Focus groups 210, 211
Forelink(s) 12, 13, 15
Forms 27, 36, 43, 44, 46, 51–53, 75, 98, 103, 116, 126, 127, 171, 174, 176, 192, 199, 221, 239
Fostering ideation 236
Foucault, Michel 43, 76
Framing 4, 26, 28, 30, 33, 35, 105, 141, 156, 240
Fuller, Buckminster 134

G

Gadamer, Hans George 121
Gender 43–46, 51, 53, 56, 57, 67, 85, 89–105, 206
Gender system 57, 90, 91
Gendered design 88–91, 94, 95, 98, 101, 102, 104
Graphic design 25, 29, 58, 82, 172, 188

H

Haraway, Donna 55, 102
HCI research 204, 207, 212
Heuristic evaluation 120
Hierarchy 90, 91, 96, 103
High-quality information 165, 167, 168, 181, 185, 187, 191, 195
Hirdman, Yvonne 90
History of UCD and HCI 29
Holistic perspective 84, 86, 204
Holistic system 109
Human centered design 69, 81, 94, 104, 112
Human centric design 94, 97, 104, 160
Human-centricity 42, 47, 61, 92, 102
Human-computer interaction (HCI) 28–32, 38, 52, 204, 207, 212
Human factors 28, 29, 32, 110

I

Ideate 24
Identities 89, 90, 99, 101
Impact 1, 3, 5–7, 28, 29, 37, 42, 47, 68, 69, 72, 77, 78, 93, 96, 108, 109, 118, 126, 148, 178, 203, 204, 229, 231
Implement 154, 166, 171, 189, 236
In situ 21, 38, 56, 58–61
Industrialization 94
Information 4, 5, 9, 14, 15, 21, 28, 29, 31, 34, 36, 38, 42, 53, 67–86, 97, 100, 102, 109, 115, 120–123, 132, 139, 148, 154, 165–170, 172, 173, 175–177, 181–185, 187, 188, 191–195, 199–201, 203, 206, 212, 216, 222–224, 226, 228, 230, 234
Information communication technology (ICT) 203
Information Design 21, 67–69, 71–74, 76, 77, 79, 82, 83, 85, 86, 165–170, 172, 173, 175, 176, 181, 185, 187, 188, 191–193, 195, 199
Innovation 1, 2, 6, 7, 10, 11, 16–18, 21, 27, 33, 41, 46, 67, 69, 89, 99, 113, 117, 123, 166, 169, 171, 195, 209, 222, 225, 235, 240
Innovative design 16
Interaction Design (IxD) 28

Interaction(s) 1, 11, 22, 25, 28, 30, 41, 46, 47, 52, 55, 58, 74, 82, 93, 96, 97, 100, 101, 110, 126, 160, 166, 204, 205, 207, 214, 216, 217, 222, 228, 233, 234, 237, 239
Interpretation 23, 26, 55, 67, 70–72, 74, 76, 97, 101, 121, 190, 222,
Intersectionality 44
Intervention(s) 140, 156, 159, 171
Iteration(s) 26, 147, 233, 236, 237

J

Jen, Natasha 27
JOURNY (JOintly Understanding Reflecting and NEgotiating strategy) 139

K

Kahneman, Daniel 48
Kelley, Tom and David 225
Kimbell, Lucy 133
Kirkham, Pat 45
Knowledge-based process 11

L

Lawson, Bryan 12, 88
Learning 11, 18, 26, 41, 45, 54, 58, 59, 62, 63, 102, 103, 105, 111, 117, 119, 130, 140, 157, 170, 173, 174, 184, 185, 187, 191, 193, 202, 203, 214, 221, 222, 226, 228, 229, 231, 233, 237–240
Learning process 11, 214, 226
Linkography 12, 17, 18
Literacy 68, 69, 72, 82, 202
Looking 1, 51, 86, 91, 97, 98, 100, 117, 131, 138, 141, 147, 154, 157, 158, 224, 233
Low-level question (LLQ) 4

M

Maintenance 47, 103, 104, 113, 172, 186, 190
Male 43, 45, 90–92, 95, 96, 98–101, 104, 221
Management Cybernetics 134, 143, 146, 151, 157
Marçal, Katrine 89, 95
Masculinity 51, 89, 90, 95, 99, 101
Material(s) 4, 41, 42, 44, 46, 47, 49, 50, 52–54, 56–58, 74, 96, 98, 99, 111, 114, 117, 120, 123, 133, 199, 210, 216, 232
Materiality norms 103
Mechanical engineering 236, 237
Memory 36, 144, 155, 178, 186
Mental images 70, 96
Mental maps 103
Mental model 70

Meso level 61
Message 48, 70–72, 75, 77, 80, 82, 85, 166, 168, 176, 203, 213, 217, 225
Metaphors 51, 76, 79, 81–83, 120, 223
Method(ology) 1, 3–7, 9–12, 16–18, 21, 23, 25, 27, 28, 30, 31, 36, 38, 47, 49, 52, 54, 57, 63, 72, 81, 83, 110–112, 114, 119, 121, 127, 132, 136–147, 149, 150, 152, 154–160, 205, 208, 209, 212, 225, 236
Micro level 61, 176
Modalities 99
Modelling 131, 132, 139, 143, 146–148, 150–157
Models of design thinking 1
Motivation 67, 68, 85, 116, 183, 184, 224, 232
Multi Criterial Decision Methods 152
Multi-layered 71, 86
Multi-methodology approaches 154
Multimodal Origami 119, 121

N

Neumier, Elizabeth 25
Norman, Donald 3, 28, 48, 221
Norms 7, 9, 42–47, 49, 51–53, 56–59, 61, 62, 88, 89, 93, 94, 98–101, 103, 104, 202
Nudging 49, 50, 52, 56–59, 63

O

Old-age 203
Operational Research 131–137, 139–143, 145, 147, 150–160
Operational Research Society 131, 132, 136, 143, 150, 152
Optimisation 133
Organization 27, 47, 113, 117, 120–123, 169, 171, 172, 189, 217
Orientation process 174
ORMS (Operations Research and Management Science) 131, 136
Osborn, Alex, F. 2

P

Package(ing) 97, 99, 169
Papanek, J. Victor 109
Paradigm 30, 31, 135, 145, 153, 156, 157, 160
Participants 24–28, 32, 38, 53, 56, 58, 68, 91, 99, 119, 121–125, 139, 140, 150, 156, 176, 180, 181, 186, 187, 191, 198, 199, 204, 205, 207–211, 213, 215–217, 230
Participatory design 50, 68, 111–113, 207, 210, 221, 240
Perception 45, 58, 68, 72–74, 76, 89, 133, 202, 203, 222, 227, 231, 240

Perception of meaning 68
Personalized guidance 185
Photo elicitation 120
Physical prototypes 232, 237, 239
Plain language 69, 76
Polanyi, Michael 111, 113
Power relations 44, 45, 53, 62, 89
Power structure 89, 92, 101
Practice 6, 26, 29, 32–34, 36, 37, 41, 42, 47, 48, 50, 52, 53, 56–58, 60, 61, 77, 88, 90, 94, 95, 110–113, 116, 118, 119, 131–134, 136–142, 144, 146, 150–152, 155–157, 159, 160, 171, 182, 183, 187, 188, 192, 222, 223, 234, 235
Practice theory 52
Pragmatism 76
Primary and secondary boundaries 117
Problem definition 24, 223
Problem solving 21, 22, 27, 34, 35, 37, 54, 132, 134, 142, 158, 193, 208
Problem Structuring 134–137, 139, 141–143, 145, 150, 152, 154, 155, 157–159
Process(es) 1–7, 9–15, 21–28, 32–34, 36–39, 41, 47, 48, 50, 52, 54–57, 59, 61, 68, 69, 71–73, 76–78, 83, 86, 92, 96–98, 102, 108, 110, 112–114, 132–133, 135, 137–142, 144, 146, 147, 150, 153–157, 159, 166, 167, 171–174, 176, 180, 187–192, 194, 198, 202–212, 214–217, 222–235, 237–239
Product design 27, 34, 45, 86, 228
Product development 86, 220, 229, 238
Protocol analysis 12, 17, 18
Prototype 226, 232–234, 236, 237, 239

Q

Question asking 3, 5

R

Radical Design 7, 8
Real world context 209
Reflection(ing) 4, 48, 51, 57, 62, 63, 77, 83, 97, 111, 126, 133, 136, 139, 140, 142, 159, 192, 206, 207, 220, 227, 231, 234, 236–238
Relationship(s) 16, 28, 35, 51, 59, 71, 72, 76, 79, 84–86, 90, 92, 93, 95, 99, 101, 104, 111, 115, 118, 123, 126, 207, 215
Research 1, 3–5, 7, 9, 11, 12, 17, 18, 21–23, 29, 32, 33, 35–37, 39, 61, 68, 69, 71, 76, 86, 92, 93, 99, 111, 112, 115–117, 119, 120, 123, 131–137, 139–143, 145, 147, 150–160, 165, 167–170, 173, 176, 195,

199, 202, 204, 207, 212, 221, 222, 224, 230, 234
Research methods 21, 32, 33, 159, 160, 176, 224
Rhetoric 27, 75, 76, 82, 83
Ritter, Friedrich 28
ROAR (Research Oriented Action Research) 139, 159

S

SAP Scences 232
Scandinavian tradition 112
Schön, Donald 49, 50, 102
Science 28, 29, 33–35, 46, 47, 50, 51, 53, 55, 67, 68, 76, 91, 92, 101, 102, 111, 131, 132, 134–136, 140, 158, 165, 172, 221
Seeing 63, 75, 97
Segment of design process 12, 18
Segregation 90, 208
Self-concept 203, 204
Self-determination 186
Self-efficacy 203, 204
Self-esteem 203, 204
Semiotic(ian) 46, 51, 52, 70, 76, 77, 101
Sense-making 50, 51, 58, 89, 225
Services 2, 6, 10, 26, 27, 30, 41, 42, 53, 67, 68, 73, 85, 101, 104, 105, 117, 165–171, 176, 178–180, 182, 184, 185, 187, 188, 191, 193, 195, 203, 205, 206, 208, 209, 210–213, 216, 217, 232, 236
Shape 29, 41, 74, 97, 98, 99, 101, 109, 213, 221, 227, 232
Simon, Herbert 134, 135, 151
Situate(d) 30, 31, 36, 42, 43, 45, 47–52, 55–63, 68, 69, 75, 76, 88, 97, 102
Situated design 41, 43, 56–63, 68, 76
Situated knowledge 52
Situatedness 42, 55, 68, 69, 77, 81, 82, 84
Situation 4, 5, 32, 48–50, 53, 54, 58, 61, 62, 67–69, 72–74, 77, 79, 84–86, 102, 104, 111, 132, 138, 140, 141, 152, 156, 167, 175, 184, 186, 193, 198, 210, 211, 223, 229, 233
Situation awareness 32, 72, 74, 77, 86
Sketch(ing) 97, 137, 138, 140, 141, 144
Social 1, 7, 9, 18, 29, 30, 32, 33, 41–47, 51–53, 55, 56, 61–63, 68, 76, 77, 81–83, 89–94, 101–105, 112, 115, 118, 131, 136, 141, 143, 152–154, 157, 158, 160, 167, 169, 170, 173, 175, 176, 178, 187–189, 191, 192, 194, 195, 198, 202–206, 209–211, 214–217, 224, 231, 232, 234, 238
Social context 175, 188, 206, 209–211, 214–216

Socio-cultural 45, 67, 75–77, 83, 156
Socio-semiotic 70
Socio-technical system 29, 133, 153, 156
SODA (Strategic Options Development Analysis) 137–139, 157, 158
Soft system methodology 136, 141–144, 156–158
SORPS (Soft Operational Research and Problem Structuring) 134–137, 142, 143, 145, 157
Space(s) 2, 8, 10, 13, 26, 30, 35, 46, 56, 67, 72, 74, 81, 83–85, 89, 93, 103, 114, 119, 124, 125, 127, 142, 146, 152, 157, 169, 173, 174, 188, 213, 226–228, 230
Speaking design 62
SPIRAL 209
Stereotypes 42, 43, 45, 56, 93, 98, 99, 104, 202–204, 206, 207, 210, 211, 213, 215, 231, 234
Strategic choice approach 143–145, 152, 157, 158
Symbols 42, 44, 46, 49, 57, 61, 62, 97, 102, 120
System 2 thinking 48, 49, 58
System dynamics 152, 154, 155
System thinking 112, 116, 117, 120, 127, 134, 136, 142, 159, 226
System(s) 2, 6, 8, 10, 18, 27–29, 32–34, 36–38, 41, 42, 48, 49, 55–58, 61, 62, 71, 72, 76–78, 88–91, 100–105, 108–110, 112, 114–121, 123–127, 131–134, 136, 139, 141–143, 146–160, 167, 169–172, 174, 178, 179, 182, 207, 208, 221, 226, 228, 232, 235, 239

T

Tangible material 114
Target group 46, 69, 75, 85, 205, 206, 227, 229, 232
Taxonomy 153
Technical toolkit 208, 213
Technology experience 206
Technology usage 204, 206, 213–216,
Test(ing) 11, 22, 26, 29, 32, 33, 37, 38, 54, 69, 72, 100, 169, 178, 180, 195, 213, 217, 222, 225, 228, 230, 232, 233, 236, 239
Thinking process 2, 4, 22, 25, 26, 33, 36, 172, 194, 227, 229, 230, 232, 234, 237, 239
Tjendra, Jeffrey 27
Tokenism 42, 55, 56

U

Unforeseen 108, 109, 116, 126
Universality 68
Usability testing 228, 232

Usable information 69
User-centered design (UCD) 21, 23, 26–33, 36–39
User centrism 239
User-experience (UX) 21, 28, 37, 58
User insight 42, 55, 56, 100, 101, 108, 112, 113, 126
User interaction 46, 101, 222
User involvement 55, 205, 209, 210
User needs 22, 23, 37, 56, 207, 209, 222, 224, 225, 232, 238
User view 189

V

Verbal information 75, 76, 78, 79, 83
Viable System Modelling 143, 147, 152, 157
Visual awareness 115
Visual communication 41, 45, 56, 90–98, 104, 117
Visual management boards 120, 121
Visual media 73
Visualize(sation) 72, 100, 122, 167, 187, 192

W

Welfare technology(ies) 165–173, 176, 178, 179, 184, 187, 188, 191–195, 198
Wicked problem 27, 41, 48, 116, 142, 143, 160, 167
Workplace 86, 104, 108, 113, 119, 153, 166, 167, 170, 181, 184, 188, 191–193

Z

Zora robot 173

Made in the USA
Monee, IL
03 May 2026